T0260000

SpringerBriefs in Computer Science

SpringerBriefs present concise summaries of cutting-edge research and practical applications across a wide spectrum of fields. Featuring compact volumes of 50 to 125 pages, the series covers a range of content from professional to academic.

Typical topics might include:

- A timely report of state-of-the art analytical techniques
- A bridge between new research results, as published in journal articles, and a contextual literature review
- A snapshot of a hot or emerging topic
- An in-depth case study or clinical example
- A presentation of core concepts that students must understand in order to make independent contributions

Briefs allow authors to present their ideas and readers to absorb them with minimal time investment. Briefs will be published as part of Springer's eBook collection, with millions of users worldwide. In addition, Briefs will be available for individual print and electronic purchase. Briefs are characterized by fast, global electronic dissemination, standard publishing contracts, easy-to-use manuscript preparation and formatting guidelines, and expedited production schedules. We aim for publication 8–12 weeks after acceptance. Both solicited and unsolicited manuscripts are considered for publication in this series.

More information about this series at http://www.springer.com/series/10028

Shu Wu · Qiang Liu · Liang Wang
Tieniu Tan

Context-Aware Collaborative Prediction

 Springer

Shu Wu
National Laboratory of Pattern
  Recognition, Institute of Automation
Chinese Academy of Sciences
Beijing
China

Liang Wang
National Laboratory of Pattern
  Recognition, Institute of Automation
Chinese Academy of Sciences
Beijing
China

Qiang Liu
National Laboratory of Pattern
  Recognition, Institute of Automation
Chinese Academy of Sciences
Beijing
China

Tieniu Tan
National Laboratory of Pattern
  Recognition, Institute of Automation
Chinese Academy of Sciences
Beijing
China

ISSN 2191-5768   ISSN 2191-5776   (electronic)
SpringerBriefs in Computer Science
ISBN 978-981-10-5372-6   ISBN 978-981-10-5373-3   (eBook)
https://doi.org/10.1007/978-981-10-5373-3

Library of Congress Control Number: 2018931488

Printed on acid-free paper

This Springer imprint is published by Springer Nature
The registered company is Springer Nature Singapore Pte Ltd.
The registered company address is: 152 Beach Road, #21-01/04 Gateway East, Singapore 189721, Singapore

Preface

Collaborative prediction becomes a fundamental technique for Internet applications, and more contextual information is available in these real scenarios. For example, the contextual information includes time, location in context-aware recommendation, system, platform, position in click rate prediction. The state-of-the-art collaborative prediction methods are based on calculating the similarity between entities and contexts, but these similarities are not always reliable. Besides, these methods are usually not able to reveal the joint characteristics among entities and contexts.

Motivated by recent works of natural language processing and representation learning, this book presents three general frameworks for context-aware modeling of collaborative prediction based on contextual representation, hierarchical representation, and context-aware recurrent neural network. This book consists of two parts. The first part introduces the theory of contextual representation providing context-aware latent vector for entities and hierarchical representation which are constructed for the joint interaction of entities and contextual information. Besides, context-aware recurrent structure is proposed for modeling contextual information and sequential information simultaneously. To provide a background to the core concepts presented, it offers an overview of contextual modeling and the background of introduced models.

The second part presents how to implement these context-aware collaborative prediction models for real tasks, such as the general recommendation, context-aware recommendation, latent collaborative retrieval, and click-through rate prediction. The proposed techniques demonstrate significant improvements over existing methods; the key determinants are the incorporated contextual modeling techniques, i.e., contextual representation, hierarchical representation, and context-aware recurrent structure. The empirical results indicate the models outperform the state-of-the-art methods of context-aware collaborative prediction and context-aware sequential prediction, on different collaborative prediction tasks.

Beijing, China   Shu Wu
December 2017

Acknowledgements

We are grateful to all members of the Center for Research on Intelligent Perception and Computing (CRIPAC), National Laboratory of Pattern Recognition (NLPR), Institute of Automation, for the constant feedback and support to this work. I am also very grateful to the students, Qiyue Yin, Weiyu Guo, Feng Yu, and Qiang Cui. They give us valuable feedback, and this work is influenced by many discussions and collaborations with them. We specially wish to thank Dr. Celine Chang, Jane Li, Shengrui Wang (UdeS), and Lifei Chen (FJNU). Their comments and critics help us to improve the quality of this book. Without their continuing help and support, this book would not have been possible. Finally, I would like to thank our family for their encouragement and support.

Research efforts summarized in this book were supported by the National Key Research and Development Program (2016YFB1001000), National Natural Science Foundation of China (61772528, 61403390, U1435221).

Shu Wu, Qiang Liu, Liang Wang, and Tieniu Tan are with the Center for Research on Intelligent Perception and Computing (CRIPAC), National Laboratory of Pattern Recognition (NLPR), Institute of Automation, Chinese Academy of Sciences (CASIA), and the University of Chinese Academy of Sciences (UCAS), Beijing, 100000, China. E-mail: {shu.wu, qiang.liu, wangliang, tnt}@nlpr.ia.ac.cn. And the corresponding author is Shu Wu.

Contents

Chapter 1
Introduction

Abstract In this chapter, we introduce the basic concepts of contextual information and collaborative prediction. Then, we introduce the scenarios of context-aware collaborative prediction and point out some limitations of the conventional methods. Finally, we introduce the tasks of collaborative prediction, on which we will compare the performance of our methods and conventional methods.

1.1 Overview

1.1.1 Contextual Information

With the rapid growth of the Internet, users are overloaded by the flooded information. Collaborative prediction systems have become an important tool to help users to select the information. Nowadays, with the enhanced ability of systems in collecting information, a great amount of contextual information is available. For example, the contextual information describes the situation of behavior, such as *location*, *time*, *weather*, *companion*. The user behavior tends to change significantly under different kinds of contexts.

The definitions of contextual information vary greatly in different research areas. In the data mining community, context is defined as those events that characterize the life of a customer and can determine the change in his/her preferences [8]. Experimental research on customer modeling shows that contextual information is helpful for predicting user-item behavior.

The survey of [1] indicates that contexts of collaborative prediction specify the contextual information associated with the application, and provides two kinds of examples which are attributes associated with users or items and attributes associated with user-item interactions. Taking recommendation systems as an example, contexts of a user, such as *gender*, *age*, and *occupation*, can profile this entity, and contexts of a user-item rating, such as *time*, *location*, *companion*, and *platform*, describe situations of this interaction. The work [10] indicates that context-aware methods

Fig. 1.1 A typical task of context-aware collaborative prediction, i.e., recommendation systems. Contextual information in recommender systems contains user contexts, item contexts, and user-item interaction contexts. User contexts or item contexts are attributes associated with the corresponding entity, and interaction contexts describe situations of the user-item interaction

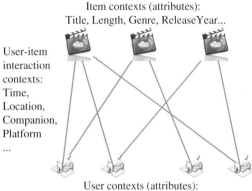

Item contexts (attributes):
Title, Length, Genre, ReleaseYear...

User-item interaction contexts: Time, Location, Companion, Platform ...

User contexts (attributes):
Gender, Age, Occupation, Label, Tweets...

are more general than attribute-aware methods [2, 14], which only consider additional information about users and items. As shown in Fig. 1.1, generally, contextual information includes interaction contexts, which describe the interaction situations, and entity contexts, which can identify user-item characteristics. Here, we focus on modeling the general contextual information associated with not only users/items but also user-item interactions.

1.1.2 Collaborative Prediction

The collaborative prediction problem is to predict the users potential preferences on unobserved items, based on their observed preferences [4, 12, 15]. This problem assumes the users and items have respective dependencies among them. As a typical task of collaborative prediction, recommendation systems attempt to predict users preferences toward different types of items (e.g., movies, books), based on previously obtained item ratings from different users.

The most extensively used approach to collaborative prediction is the factorization model. The basic assumption of these models is that there are a small number of latent factors which affect user's preferences toward the items [12]. For instance, based on a matrix factorization model, a user's preference on an item is computed by the coefficients of user interests and the values of item properties.

Nowadays, collaborative prediction plays an important role in numerous real-world applications, e.g., recommender systems, information retrieval, and social network analysis. With the rapid growth of Web applications, there are more and more entities and contexts. For example, there are three entities in tag recommendation (i.e., user, item, and tag) [9, 11] and three or more entities and contexts in context-aware recommendation (i.e., user, query, and several contexts) [3, 5, 10, 13]. Moreover, in some probability prediction applications, e.g., click-through rate prediction [6, 7, 16], there are even more entities and contexts in an interaction situation (e.g., device, Web site, application, advertisement, and position). Accordingly, it is necessary to model joint characteristics of entities and contexts in interaction scenarios and to collaborative predict interaction among multiple entities and contexts.

1.1.3 Context-Aware Collaborative Prediction

Due to the fundamental effect of contextual information in collaborative prediction, many context modeling methods have been developed. These methods fall into contextual filtering and contextual modeling. The contextual filtering approach uses contexts to select data and adjust the results, and contextual modeling directly takes contexts into the model construction. As a popular contextual modeling method, factorization-based contextual modeling [3, 10] incorporates contextual information in a factorization model via treating contexts as one or several additional dimensions similar to the dimensions of the user and the item. These methods compute the relevance between contexts and entities, but such kind of relevance is not always reasonable [13]. For example, it is not intuitive that a user is more relevant to *weekday* than *weekend*. In 2014, Shi et al. propose a novel $CARS^2$ model [13] which provides each user/item with not only a latent vector but also an effective context-aware representation. However, using a distinct vector to represent contexts of each interaction, $CARS^2$ has the problem in confronting with abundant contextual information in real applications. Besides, since $CARS^2$ can only model the categorical context, the numerical context should be categorized at first.

On the other hand, with multiple entities and contexts, factorization methods make collaborative prediction of entity and context interactions based on calculating the similarities between all the entities and contexts. This does not conform to situations of various applications. For example, in latent collaborative retrieval, factorization methods compute the similarity among user, query, and document, but the similarity between the user and the query does not contribute to the document selection. These methods are not able to reveal the joint characteristics of entities and contexts in the interaction. They only calculate the similarity based on corresponding dimensional values in the latent vectors, which does not allow rich high-order calculation among values of different dimensions.

1.1.4 Tasks

This book will empirically investigate the performance of COT, HIR, CA-RNN, and other methods on different collaborative prediction tasks. These tasks are general recommendation, context-aware recommendation, latent collaborative retrieval, click-through rate prediction, and context-aware sequential prediction. They are representative tasks of context-aware collaborative prediction and contain different number of entities and contexts. The first two tasks are recommendation systems with different number of contexts. The third and fourth tasks are with abundant contextual information. The final one, context-aware sequential prediction, contains not only common contexts but also sequential contexts. The experimental results on real datasets show that COT, HIR, and CA-RNN outperform the state-of-the-art context-aware collaborative prediction and context-aware sequential prediction, respectively.

The impact of the dimensionality of latent representations is analyzed, and we also examine the interacting order of entities and contexts in context-aware recommendation and latent collaborative retrieval. Then, we visualize the representations in latent collaborative retrieval to find interesting observations on context representations. We find some interesting observations when visualizing the representations of entities, contexts, and their interaction. For context-aware sequential prediction, we conduct experiments on comparing different input contexts and different aggregation methods of input contexts.

1.2 Book Structure

The rest of this book is organized as follows. Chapter 2 introduces the general ways to integrate contexts with collaborative prediction, i.e., contextual filtering and contextual modeling. Then, we review some related tasks and methods of collaborative prediction. Chapter 3 presents the contextual operation for context-aware modeling. It includes the basic notations and fundamental concepts of contextual operation, contextual operating tensor model, the process of parameter inference, and optimization algorithm. Chapter 4 describes the hierarchical interaction representation model, which models the interaction as representation among different entities and contexts. For modeling sequential information and contextual information, a sequential prediction model based on context-aware recurrent structure is presented in Chap. 5. Chapter 6 describes the existing methods of different prediction tasks and empirically investigates the performance of our methods and the representative methods on these tasks.

References

1. Adomavicius, G., Tuzhilin, A.: Context-aware recommender systems. In: Recommender Systems Handbook, pp. 217–253. Springer, Berlin (2011)
2. Agarwal, D., Chen, B.C.: Regression-based latent factor models. In: Proceedings of the 15th ACM SIGKDD International Conference on Knowledge Discovery and Data Mining, pp. 19–28. ACM, New York (2009)
3. Karatzoglou, A., Amatriain, X., Baltrunas, L., Oliver, N.: Multiverse recommendation: n-dimensional tensor factorization for context-aware collaborative filtering. In: Proceedings of the fourth ACM conference on Recommender systems, pp. 79–86. ACM, New York (2010)

4. Kim, Y., Choi, S.: Bayesian binomial mixture model for collaborative prediction with non-random missing data. In: Eighth ACM Conference on Recommender Systems, RecSys '14, Foster City, Silicon Valley, CA, USA, pp. 201–208, 06–10 Oct 2014

5. Liu, Q., Wu, S., Wang, L.: Cot: Contextual operating tensor for context-aware recommender systems. In: Proceedings of the 29th International AAAI Conference on Artificial Intelligence, pp. 203–209. AAAI, California (2015)

6. McMahan, H.B., Holt, G., Sculley, D., Young, M., Ebner, D., Grady, J., Nie, L., Phillips, T., Davydov, E., Golovin, D., et al.: Ad click prediction: a view from the trenches. In: Proceedings of the 19th ACM SIGKDD international conference on Knowledge discovery and data mining, pp. 1222–1230. ACM, New York (2013)

7. Oentaryo, R.J., Lim, E.P., Low, J.W., Lo, D., Finegold, M.: Predicting response in mobile advertising with hierarchical importance-aware factorization machine. In: Proceedings of the 7th ACM international conference on Web search and data mining, pp. 123–132. ACM, New York (2014)

8. Palmisano, C., Tuzhilin, A., Gorgoglione, M.: Using context to improve predictive modeling of customers in personalization applications. IEEE Trans. Knowl. Data Eng. **20**(11), 1535–1549 (2008)

9. Rendle, S., Balby Marinho, L., Nanopoulos, A., Schmidt-Thieme, L.: Learning optimal ranking with tensor factorization for tag recommendation. In: Proceedings of the 15th ACM SIGKDD International Conference on Knowledge Discovery and Data Mining, pp. 727–736. ACM, New York (2009)

10. Rendle, S., Gantner, Z., Freudenthaler, C., Schmidt-Thieme, L.: Fast context-aware recommendations with factorization machines. In: Proceedings of the 34th international ACM SIGIR conference on Research and development in Information Retrieval, pp. 635–644. ACM, New York (2011)

11. Rendle, S., Schmidt-Thieme, L.: Pairwise interaction tensor factorization for personalized tag recommendation. In: Proceedings of the Third ACM International Conference on Web Search and Data Mining, pp. 81–90. ACM, New York (2010)

12. Rish, I., Tesauro, G.: Active collaborative prediction with maximum margin matrix factorization. In: International Symposium on Artificial Intelligence and Mathematics, ISAIM 2008, Fort Lauderdale, Florida, USA, 2–4 Jan 2008

13. Shi, Y., Karatzoglou, A., Baltrunas, L., Larson, M., Hanjalic, A.: Cars2: Learning context-aware representations for context-aware recommendations. In: Proceedings of the 23rd ACM International Conference on Conference on Information and Knowledge Management, pp. 291–300. ACM, New York (2014)

14. Stern, D.H., Herbrich, R., Graepel, T.: Matchbox: Large scale online bayesian recommendations. In: Proceedings of the 18th International Conference on World Wide Web, pp. 111–120. International World Wide Web Conferences Steering Committee, Geneva (2009)

15. Xu, M., Zhu, J., Zhang, B.: Nonparametric max-margin matrix factorization for collaborative prediction. In: 26th Annual Conference on Neural Information Processing Systems, pp. 64–72. (2012)

16. Yan, L., Li, W.J., Xue, G.R., Han, D.: Coupled group lasso for web-scale ctr prediction in display advertising. In: Proceedings of the 31th International Conference on Machine Learning, pp. 802–810. ACM, New York (2014)

Chapter 2
Context-Aware Collaborative Prediction

Abstract Context-aware collaborative prediction takes contextual information into consideration when modeling user preferences and predicting user behaviors. There are two general ways to integrate contexts with collaborative prediction: contextual filtering and contextual modeling. Contextual filtering uses contexts to select data and adjust the result, while contextual modeling takes contexts into the model construction. Currently, the most effective context-aware collaborative prediction algorithms are based on the contextual modeling approach, which generates contextual representations or context-aware representations. This chapter reviews some related tasks of collaborative prediction, such as conventional recommendation, sequential recommendation, and multi-domain relation prediction. In addition, it also introduces some recent works on representation learning and methods of specific applications, such as context-aware recommendation, latent collaborative retrieval, tag recommendation, and click-through rate prediction.

2.1 Context Modeling Methods

Contextual information has been proved to be useful for collaborative prediction [2, 41], and various context-aware collaborative prediction methods have been developed. According to the survey of [2], these methods can be categorized into contextual filtering and context modeling.

2.1.1 Contextual Filtering

Employing the pre-filtering or post-filtering strategy, conventional contextual filtering methods [1, 5, 42] utilize the contextual information to drive data selection or adjust the resulting set. For instance, Li et al. view a context as a dynamic feature of items and filter out the items that do not match a specific context [24]. Some works [30, 71] have applied tree-based partition with matrix factorization, which

© The Author(s) 2017
S. Wu et al., *Context-Aware Collaborative Prediction*, SpringerBriefs in Computer Science, https://doi.org/10.1007/978-981-10-5373-3_2

also fall into the pre-filtering category. The work of [70] combines the user and item splitting in the dataset to boost context-aware collaborative prediction and performs an empirical comparison of these three contextual filtering approaches on multiple datasets. To deal with contexts and social network in recommender systems, Liu et al. propose SoCo [30], which splits contexts and performs general matrix factorization only on the leaf nodes of decision trees. These pre-filtering and post-filtering methods may work in practice, but they require supervision and fine-tuning in all steps of collaborative prediction [47].

2.1.2 Contextual Modeling

The context modeling methods, using the contextual information directly in the model, have become popular recently. Most of these methods are based on factorization methods or contextual representation methods. The first approach focuses on integrating the contextual information with the user-item rating matrix and constructing factorization models. The contextual representation methods generate the vector representations of contexts or provide users/items with context-aware representations.

2.1.2.1 Factorization Methods

Factorization methods of the contextual modeling approach employ factorization models to compute collaborative prediction based on the user-item preference relations and contextual information. Multiverse recommendation [17] uses tensor factorization to model n-dimensional contextual information and is factorized with Tucker decomposition [58]. Multiverse recommendation has proved performing better than the conventional contextual pre-filtering and post-filtering models. The Tensor Factorization for MAP maximization (TFMAP) model [51] uses tensor factorization and Mean Average Precision (MAP) objective to model implicit feedback data with contextual information. As an extended model of tensor factorization, factorization machine (FM) can model a wide variety of contextual information by specifying contextual information as the input dimensions and provide context-aware predictions [47]. This method can handle different kinds of contextual information and factorizes pair-wise context relation through generating feature vectors in a proper way. For effective tensor factorization, MRTensorCube [18] is designed based on the MapReduce computation framework and supports efficient context recognition. The work on Heterogeneous Matrix Factorization (HeteroMF) [15] generates context-specific latent vectors of entities using a context-dependent transfer matrix and the original latent vectors of entities. However, since these methods treat contexts as one or several dimensions as those of the user and item, the relation between a context value and an entity is not intuitive and has difficulty in explanation.

2.1.2.2 Contextual Representation

Based on neighborhood-based methods, contextual contents of [31] are combined via a function to produce the context influence factor. Then, combining this context influence factor with each latent factor can generate context-aware latent representations. The context-aware embedding method [61] incorporates the contextual information with users rating behavior, based on the distance functions of the points representing user and item in embedding space. This method allows visualization as an analytics tool for context-aware recommendation. Recently, Shi et al. propose a novel $CARS^2$ [50] model which provides each user/item with not only a latent factor but also a context-aware representation. Similar to HeteroMF [15], $CARS^2$ provides the contextual information of each interaction with a distinct vector. But these methods are not suitable for numerical contexts and abundant contexts in real-world applications. The contextual operating tensor (COT) models [28, 62] and the hierarchical interaction representation (HIR) model [27] introduced in this book also fall into the contextual representation category. COT represents contexts as latent vectors and captures the common semantic effects of contexts by using contextual operating tensor. For the contextual information of each user-item interaction, the contextual operation can be modeled by multiplying the operating tensor with latent vectors of contexts. HIR models the mutual action among different entities and contexts as an interaction representation by using tensor multiplication. This method is performed iteratively to construct a hierarchical structure among all entities.

2.2 Methods of Collaborative Prediction

This section introduces some related methods which have been used in different tasks of collaborative prediction. Conventional recommendation and sequential prediction are very common tasks, which usually employ factorization and Markov chain-based methods. Most methods of multi-domain relation prediction are also based on factorization. In addition, this section introduces some recent works on representation learning and methods in specific applications, i.e., context-aware recommendation, latent collaborative retrieval, and click-through rate prediction.

2.2.1 Recommender Algorithms

Matrix factorization methods (MF) [21, 22, 37] have become the state-of-the-art approach to recommender systems, due to their attractive accuracy and scalability. The basic objective of MF is to factorize a user-item rating matrix into two low-rank matrices, each of which represents the latent factors of users or items. With the multiplication of two factorized matrices, the original matrix can be reconstructed, and rating predictions are obtained accordingly. Matrix factorization-based methods have been extensively studied, e.g., SVD++ [19], tensor factorization [63], and factorization machine (FM) [43]. SVD++ combines neighborhood models with latent factor models in one prediction function. Tensor factorization (TF) [63] extends MF from two dimensions to three or even higher dimensions. FM gives a further extension of TF by modeling all interactions between pairs of entities. Both TF and FM are successfully used in context-aware recommendation [17, 47] and tag recommendation [44, 48, 57]. Aiming to maximize the probability difference between positive items and negative items, Bayesian Personalized Ranking (BPR) [45] is a commonly used method for modeling implicit feedback of users. It has become a state-of-the-art framework for personalized ranking. The attribute-aware recommender systems [3, 56] extend the conventional MF model to handle user and item attributes, and this attribute-aware MF is another important direction of MF extensions.

2.2.2 Sequential Prediction

Time-aware neighborhood models [11, 23, 26] may be the most natural method for modeling sequences, which employ neighborhood-based algorithms to capture temporal effects via giving more relevance to recent observations and less to past observations. Besides, there are some MF-based methods which are designed for the time factor of sequences. TimeSVD++ [20], which learns time-aware representations for users and items, is one of the most effective models for time-aware recommendation. Tensor factorization (TF) [4, 63] treats time intervals as another dimension to the rating matrix and generates latent vectors of time intervals via factorization to capture the underlying properties in historical sequences. But factorization-based models have difficulties in generating latent representations for time intervals which has never or seldom appeared in the training data. Using frequent pattern mining, sequential pattern-based methods [13, 38] seek sequential patterns that occur most frequently to predict the future. Sequential pattern-based methods are unable to reveal the underlying properties in users' historical sequences and are time consuming in large-scale datasets.

The MC-based methods are widely used models for sequential applications [65]. Via factorization of the probability transition matrix, Factorizing Personalized Markov Chain (FPMC) [46] can provide more accurate prediction for each sequence. FPMC is also extended by using user group [39] or incorporating location constraint [10]. Recently, some factors of human brain have been added into MC-based methods, including interest-forgetting curve [9] and dynamics of boredom [16]. The main drawback of MC-based models is the independent combination of the past components, which lies in a strong independence assumption and confines the prediction accuracy. MC-based methods are then extended by representation learning. Hierarchical Representation Model (HRM) [59] learns the representation of behaviors in the last record and predicts behaviors for the next record. And Personalized Ranking Metric Embedding (PRME) [12] learns embeddings of users according to the location distance.

Recently, a few prediction models, especially language models, are proposed based on neural networks. Among them, recurrent neural networks (RNNs) become the most popular one. RNN-based models have been successfully used in modeling sentences [33–35]. RNN also brings satisfying results for sequential click prediction for sponsored search [69], location prediction [29], and next basket recommendation [67]. Backpropagation Through Time (BPTT) [49] is usually used for learning of RNN models.

2.2.3 Multi-domain Relation Prediction

Multi-domain relation prediction can also be used for the context-aware collaborative prediction. For relation learning in multiple domains [25, 52, 66], Collective Matrix Factorization (CMF) factorizes the user-item rating matrix in each domain, and latent vectors of users/items are shared among these domains. Then, Zhang et al. [68] treat user attributes as priors for user latent vectors and employ a transfer matrix to generate latent vectors from the general ones. Similarly, Jamali et al. propose Heterogeneous Matrix Factorization (HeteroMF), which generates context-specific latent vectors using a general latent vector for the entity and context-dependent transfer matrices [15]. For the context-aware collaborative prediction, with a transfer matrix for contexts in each interaction event, these methods have to estimate numerous matrices for a large amount of contextual information.

2.2.4 Representation Learning

This part introduces several most significant works of representation learning in natural language processing (NLP), which motivate the COT and HIR models. For continuous vectors of words, the neural network language model [8] is a popular and classic work, which learns a vector representation of each word. Mikolov et al. [36] propose neural net language models for computing continuous vector representations of words and provide the tool $word2vec$ for an efficient implementation. For sentence sentiment detection, the work [6] introduces a presentation of adjective-noun phrase, where a noun has semantic information as a latent vector and an adjective has semantic operation on nouns as an operating matrix, then the adjective-noun composition can be represented by multiplying the adjective matrix with the noun vector. Further, Socher et al. propose a model [54] in which each word or longer phrase has a Matrix-Vector representation. The vector captures the meaning of the constituent, and the matrix describes how it modifies the meaning of the other combined word. Since each word has a Matrix-Vector representation, the number of parameters becomes very large with an increasing size of vocabulary. Then, Socher et al. [55] propose a global tensor-based composition function for all combinations and improve the performance of sentence sentiment detection over the Matrix-Vector representation [54].

2.2.5 Application-Specific Methods

Context-aware recommendation contains entities and contexts (i.e., user, query, and several contexts), and the context modeling approach has made significant improvement of this task. Recent works integrate contextual information with user-item ratings and build factorization models. Incorporating tensor factorization, multiverse recommendation [17] represents the user-item rating matrix with contextual information as a user-item-context rating tensor, which is factorized with Tucker decomposition [58]. And FM is easily applicable to a wide variety of contexts by specifying only the input data [47]. Random decision trees have also been applied here, in which contexts are split and general matrix factorization is performed on each leaf [30]. Furthermore, the work of [68] considers user attributes as priors for user latent vectors, and a transfer matrix is used to generate latent vectors from the original ones.

Latent collaborative retrieval (LCR) [14, 60] has been proposed for modeling the interaction among user, query, and document. LCR represents a user as a matrix and generates the joint representation of user and query via multiplication. Then the prediction is given based on the inner product of latent representation of query and joint representation of user and query. Modeling the interaction among user, item, and tag, tag recommendation can also be viewed as a LCR problem, in which a user retrieves tags with an item. Treating user, item, and tag as three dimensions of a tensor, TF has been successfully used in this problem [44, 57]. As an extended version of TF, FM is also applied in tag recommendation and has become one of the state-of-the-art methods [48].

In the task of click-through rate (CTR) prediction, there are many entities and contexts, such as user, device, Web site, application, advertisement, position. Due to its ease of implementation and promising performance, logistic regression (LR) has been widely used for CTR prediction, especially in industrial systems [32, 64]. LR can be used in the prediction problem of entity and context interaction with one-hot representation of inputs. LR or other classifiers have difficulties in discovering latent relation among inputs, and the final prediction of LR could be viewed as the sum of the biases of all the entities and contexts. Recently, as a representative factorization method, FM is applied in CTR prediction and brings a great improvement [40]. Predicting interactions of entities and contexts based on calculating the similarities, FM is not able to reveal the joint characteristics of interacting entities and contexts.

These methods mentioned above achieve delightful results in respective tasks of context-aware collaborative prediction. But most of them are unable to be extended effectively for other tasks of collaborative prediction or have difficulty in dealing with more entities and contexts.

2.3   Contextual Operation

Contextual operating tensor model, named COT, is motivated by the recent work of semantic compositionality in natural language processing (NLP). Continuous vector representations of words have a long history in NLP and become even popular since Mikolov et al. [36] provides an efficient implementation *word2vec*. Inspired by the powerful ability in describing latent properties of words, in the collaborative prediction, using a vector representation of each context value seems a good solution to examine the effect of contexts on user-item interactions. This distributed representation inferred from all contexts has more powerful ability in illustrating the operation properties of contexts.

In the research direction of sentence sentiment detection, a noun has semantic information as a latent vector, and an adjective has semantic operation on nouns as an operating matrix [6, 54]. For instance, in the phrase "excellent product," the noun "product" is represented by a latent vector, and the adjective "excellent" is associated with a semantic operating matrix which can operate the noun vector of "product." Thus, multiplying the operating matrix with the latent vector, the phrase "excellent product" has a new latent vector. We assume that contexts in the collaborative prediction have a similar property of adjectives and can operate latent characteristics of users and items. Then, new latent representations of entities can show not only characteristics of original entities but also new properties under a specific contextual situation. For instance, a man has his original latent interests. When this man is with children, this *companion* context operates his latent interests and he may like to watch cartoons with these children. Besides, in real collaborative prediction systems, some contexts have very similar effects. For instance, both *weekend* and being at *home* may make you prefer to read novels. Inspired by [55] in simplifying the Matrix-Vector operation, we use contextual operating tensors to capture the common effects of contexts.

2.4   Hierarchical Representation

Recent studies on representation learning [7] in different research areas give us great inspiration in constructing models for context-aware collaborative prediction. In recent works of natural language processing, using tensor multiplication, neural tensor networks are successfully used in learning the representation of two entities for knowledge base completion [53] and the representation of a sentence for semantic compositionality [55]. Tensor multiplication is also used in modeling contextual representation for context-aware recommendation [28, 50, 62].

We introduce a hierarchical interaction representation (HIR) model for predicting interaction among entities and contexts. In HIR, each entity or context is represented as a latent vector. And we use three-dimensional tensor multiplication to capture the joint characteristics of entities and contexts. Tensor multiplication allows high-order calculation between these entities and contexts. Moreover, rather than achieving a score of the interaction in conventional methods, HIR generates a joint representation, which allows interacting with the next entity or context and can be implemented for numerous applications. In interaction scenario with n entities and contexts, this tensor multiplying procedure can be performed iteratively, which forms a hierarchical architecture with n layers. The ith layer presents the joint representation of the first i entities and contexts in this scenario. The nth layer is the final representation of all the entities and contexts, which describes their joint characteristics. With such a hierarchical structure, we can also add new entities or contexts directly without retraining the former representations. Based on the final representation, we could select learning methods according to different application tasks for collaborative prediction (i.e., linear regression for regression tasks, pair-wise ranking method for ranking tasks, and logistic regression for classification tasks).

References

1. Adomavicius, G., Sankaranarayanan, R., Sen, S., Tuzhilin, A.: Incorporating contextual information in recommender systems using a multidimensional approach. ACM Trans. Inf. Syst. (TOIS) **23**(1), 103–145 (2005)
2. Adomavicius, G., Tuzhilin, A.: Context-aware recommender systems. In: Recommender Systems Handbook, pp. 217–253. Springer, Berlin (2011)
3. Agarwal, D., Chen, B.C.: Regression-based latent factor models. In: Proceedings of the 15th ACM SIGKDD International Conference on Knowledge Discovery and Data Mining, pp. 19–28. ACM, New York (2009)
4. Bahadori, M.T., Yu, Q.R., Liu, Y.: Fast multivariate spatio-temporal analysis via low rank tensor learning. In: NIPS, pp. 3491–3499. (2014)
5. Baltrunas, L., Ricci, F.: Context-based splitting of item ratings in collaborative filtering. In: Proceedings of the third ACM conference on Recommender systems, pp. 245–248. ACM, New York (2009)
6. Baroni, M., Zamparelli, R.: Nouns are vectors, adjectives are matrices: representing adjective-noun constructions in semantic space. In: Proceedings of the 2010 Conference on Empirical Methods in Natural Language Processing, pp. 1183–1193. Association for Computational Linguistics (2010)
7. Bengio, Y., Courville, A., Vincent, P.: Representation learning: a review and new perspectives. IEEE Trans. Pattern Anal. Mach. Intell. **35**(8), 1798–1828 (2013)
8. Bengio, Y., Ducharme, R., Vincent, P., Jauvin, C.: A neural probabilistic language model. J. Mach. Learn. Res. **3**, 1137–1155 (2003)
9. Chen, J., Wang, C., Wang, J.: A personalized interest-forgetting markov model for recommendations. In: AAAI, pp. 16–22. (2015)
10. Cheng, C., Yang, H., Lyu, M.R., King, I.: Where you like to go next: successive point-of-interest recommendation. In: IJCAI, pp. 2605–2611. (2013)
11. Ding, Y., Li, X.: Time weight collaborative filtering. In: CIKM, pp. 485–492. (2005)
12. Feng, S., Li, X., Zeng, Y., Cong, G., Chee, Y.M., Yuan, Q.: Personalized ranking metric embedding for next new poi recommendation. In: IJCAI, pp. 2069–2075. (2015)
13. Hariri, N., Mobasher, B., Burke, R.: Context-aware music recommendation based on latenttopic sequential patterns. In: RecSys, pp. 131–138. (2012)
14. Hsiao, K.J., Kulesza, A., Hero, A.: Social collaborative retrieval. In: Proceedings of the 7th ACM International Conference on Web Search and Data Mining, pp. 293–302. ACM, New York (2014)

15. Jamali, M., Lakshmanan, L.: Heteromf: recommendation in heterogeneous information networks using context dependent factor models. In: Proceedings of the 22nd international conference on World Wide Web, pp. 643–654. International World Wide Web Conferences Steering Committee, Geneva (2013)
16. Kapoor, K., Subbian, K., Srivastava, J., Schrater, P.: Just in time recommendations: Modeling the dynamics of boredom in activity streams. In: WSDM, pp. 233–242. (2015)
17. Karatzoglou, A., Amatriain, X., Baltrunas, L., Oliver, N.: Multiverse recommendation: n-dimensional tensor factorization for context-aware collaborative filtering. In: Proceedings of the fourth ACM conference on Recommender systems, pp. 79–86. ACM, New York (2010)
18. Kim, S., Lee, S., Kim, J., Yoon, Y.I.: Mrtensorcube: tensor factorization with data reduction for context-aware recommendations. J. Supercomput. (2017)
19. Koren, Y.: Factorization meets the neighborhood: a multifaceted collaborative filtering model. In: Proceedings of the 14th ACM SIGKDD international conference on Knowledge discovery and data mining, pp. 426–434. ACM, New York (2008)
20. Koren, Y.: Collaborative filtering with temporal dynamics. Commun. ACM **53**(4), 89–97 (2010)
21. Koren, Y., Bell, R.: Advances in collaborative filtering. In: Recommender Systems Handbook, pp. 145–186. Springer, Berlin (2011)
22. Koren, Y., Bell, R., Volinsky, C.: Matrix factorization techniques for recommender systems. Computer **42**(8), 30–37 (2009)
23. Lathia, N., Hailes, S., Capra, L.: Temporal collaborative filtering with adaptive neighbourhoods. In: SIGIR, pp. 796–797. (2009)
24. Li, Y., Nie, J., Zhang, Y., Wang, B., Yan, B., Weng, F.: Contextual recommendation based on text mining. In: Proceedings of the 23rd International Conference on Computational Linguistics, pp. 692–700. Association for Computational Linguistics (2010)
25. Lippert, C., Weber, S.H., Huang, Y., Tresp, V., Schubert, M., Kriegel, H.P.: Relation prediction in multi-relational domains using matrix factorization. In: Workshops of Neural Information Processing Systems: Structured Input-Structured Output. (2008)
26. Liu, N.N., Zhao, M., Xiang, E., Yang, Q.: Online evolutionary collaborative filtering. In: RecSys, pp. 95–102. (2010)
27. Liu, Q., Wu, S., Wang, L.: Collaborative prediction for multi-entity interaction with hierarchical representation. In: CIKM, pp. 613–622. (2015)
28. Liu, Q., Wu, S., Wang, L.: Cot: Contextual operating tensor for context-aware recommender systems. In: Proceedings of the 29th International AAAI Conference on Artificial Intelligence, pp. 203–209. AAAI, California (2015)
29. Liu, Q., Wu, S., Wang, L., Tan, T.: Predicting the next location: A recurrent model with spatial and temporal contexts. In: AAAI, pp. 194–200. (2016)
30. Liu, X., Aberer, K.: Soco: a social network aided context-aware recommender system. In: Proceedings of the 22nd international conference on World Wide Web, pp. 781–802. International World Wide Web Conferences Steering Committee (2013)
31. Liu, X., Wu, W.: Learning context-aware latent representations for context-aware collaborative filtering. In: Proceedings of the 38th International ACM SIGIR Conference on Research and Development in Information Retrieval, Santiago, Chile, pp. 887–890, 9–13 Aug, 2015
32. McMahan, H.B., Holt, G., Sculley, D., Young, M., Ebner, D., Grady, J., Nie, L., Phillips, T., Davydov, E., Golovin, D., et al.: Ad click prediction: a view from the trenches. In: Proceedings of the 19th ACM SIGKDD international conference on Knowledge discovery and data mining, pp. 1222–1230. ACM, New York (2013)
33. Mikolov, T., Karafiát, M., Burget, L., Cernockỳ, J., Khudanpur, S.: Recurrent neural network based language model. In: INTERSPEECH, pp. 1045–1048. (2010)
34. Mikolov, T., Kombrink, S., Burget, L., Cernocky, J.H., Khudanpur, S.: Extensions of recurrent neural network language model. In: ICASSP, pp. 5528–5531. (2011)
35. Mikolov, T., Kombrink, S., Deoras, A., Burget, L., Cernocky, J.: Rnnlm-recurrent neural network language modeling toolkit. In: ASRU Workshop, pp. 196–201. (2011)
36. Mikolov, T., Sutskever, I., Chen, K., Corrado, G., Dean, J.: Distributed representations of words and phrases and their compositionality. In: Proceedings on Neural Information Processing Systems. (2013)

37. Mnih, A., Salakhutdinov, R.: Probabilistic matrix factorization. In: Proceedings on Neural Information Processing systems. (2007)
38. Mobasher, B., Dai, H., Luo, T., Nakagawa, M.: Using sequential and non-sequential patterns in predictive web usage mining tasks. In: ICDM, pp. 669–672. (2002)
39. Natarajan, N., Shin, D., Dhillon, I.S.: Which app will you use next?: Collaborative filtering with interactional context. In: RecSys, pp. 201–208. (2013)
40. Oentaryo, R.J., Lim, E.P., Low, J.W., Lo, D., Finegold, M.: Predicting response in mobile advertising with hierarchical importance-aware factorization machine. In: Proceedings of the 7th ACM international conference on Web search and data mining, pp. 123–132. ACM, New York (2014)
41. Palmisano, C., Tuzhilin, A., Gorgoglione, M.: Using context to improve predictive modeling of customers in personalization applications. IEEE Trans. Knowl. Data Eng. **20**(11), 1535–1549 (2008)
42. Panniello, U., Tuzhilin, A., Gorgoglione, M., Palmisano, C., Pedone, A.: Experimental comparison of pre-vs. post-filtering approaches in context-aware recommender systems. In: Proceedings of the third ACM conference on Recommender systems, pp. 265–268. ACM, New York (2009)
43. Rendle, S.: Factorization machines with libfm. ACM Trans. Intell. Syst. Technol. (TIST) **3**(3), 57 (2012)
44. Rendle, S., Balby Marinho, L., Nanopoulos, A., Schmidt-Thieme, L.: Learning optimal ranking with tensor factorization for tag recommendation. In: Proceedings of the 15th ACM SIGKDD International Conference on Knowledge Discovery and Data Mining, pp. 727–736. ACM, New York (2009)
45. Rendle, S., Freudenthaler, C., Gantner, Z., Schmidt-Thieme, L.: Bpr: Bayesian personalized ranking from implicit feedback. In: Proceedings of the 25th Conference on Uncertainty in Artificial Intelligence, pp. 452–461. AUAI Press (2009)
46. Rendle, S., Freudenthaler, C., Schmidt-Thieme, L.: Factorizing personalized markov chains for next-basket recommendation. In: WWW, pp. 811–820. (2010)
47. Rendle, S., Gantner, Z., Freudenthaler, C., Schmidt-Thieme, L.: Fast context-aware recommendations with factorization machines. In: Proceedings of the 34th international ACM SIGIR conference on Research and development in Information Retrieval, pp. 635–644. ACM, New York (2011)
48. Rendle, S., Schmidt-Thieme, L.: Pairwise interaction tensor factorization for personalized tag recommendation. In: Proceedings of the Third ACM International Conference on Web Search and Data Mining, pp. 81–90. ACM, New York (2010)
49. Rumelhart, D.E., Hinton, G.E., Williams, R.J.: Learning representations by back-propagating errors. Cogn. Model. **5**, 3 (1988)
50. Shi, Y., Karatzoglou, A., Baltrunas, L., Larson, M., Hanjalic, A.: Cars2: Learning context-aware representations for context-aware recommendations. In: Proceedings of the 23rd ACM International Conference on Conference on Information and Knowledge Management, pp. 291–300. ACM, New York (2014)
51. Shi, Y., Karatzoglou, A., Baltrunas, L., Larson, M., Hanjalic, A., Oliver, N.: Tfmap: Optimizing map for top-n context-aware recommendation. In: SIGIR, pp. 155–164. (2012)
52. Singh, A.P., Gordon, G.J.: Relational learning via collective matrix factorization. In: Proceedings of the 14th ACM SIGKDD international conference on Knowledge discovery and data mining, pp. 650–658. ACM, New York (2008)
53. Socher, R., Chen, D., Manning, C.D., Ng, A.: Reasoning with neural tensor networks for knowledge base completion. In: Advances in Neural Information Processing Systems, pp. 926–934. (2013)
54. Socher, R., Huval, B., Manning, C.D., Ng, A.Y.: Semantic compositionality through recursive matrix-vector spaces. In: Proceedings of the 2012 Joint Conference on Empirical Methods in Natural Language Processing and Computational Natural Language Learning, pp. 1201–1211. Association for Computational Linguistics (2012)

55. Socher, R., Perelygin, A., Wu, J.Y., Chuang, J., Manning, C.D., Ng, A.Y., Potts, C.: Recursive deep models for semantic compositionality over a sentiment treebank. In: Proceedings of the Conference on Empirical Methods in Natural Language Processing (EMNLP), pp. 1631–1642. Association for Computational Linguistics (2013)
56. Stern, D.H., Herbrich, R., Graepel, T.: Matchbox: Large scale online bayesian recommendations. In: Proceedings of the 18th International Conference on World Wide Web, pp. 111–120. International World Wide Web Conferences Steering Committee (2009)
57. Symeonidis, P., Nanopoulos, A., Manolopoulos, Y.: Tag recommendations based on tensor dimensionality reduction. In: Proceedings of the Second ACM Conference on Recommender Systems, pp. 43–50. ACM, New York (2008)
58. Tucker, L.R.: Some mathematical notes on three-mode factor analysis. Psychometrika **31**(3), 279–311 (1966)
59. Wang, P., Guo, J., Lan, Y., Xu, J., Wan, S., Cheng, X.: Learning hierarchical representation model for next basket recommendation. In: SIGIR, pp. 403–412. ACM, New York (2015)
60. Weston, J., Wang, C., Weiss, R., Berenzweig, A.: Latent collaborative retrieval. In: Proceedings of the 29th International Conference on Machine Learning, pp. 9–16. ACM, New York (2012)
61. Wu, K.K., Liu, P., Meng, H.M., Yam, Y.: An embedding approach for context-aware collaborative recommendation and visualization. In: 2016 IEEE International Conference on Systems, Man, and Cybernetics, SMC 2016, Budapest, Hungary, pp. 3457–3462, 9–12 Oct, 2016. https://doi.org/10.1109/SMC.2016.7844768
62. Wu, S., Liu, Q., Wang, L., Tan, T.: Contextual operation for recommender systems. IEEE TKDE **28**, 2000–2012 (2016)
63. Xiong, L., Chen, X., Huang, T.K., Schneider, J.G., Carbonell, J.G.: Temporal collaborative filtering with bayesian probabilistic tensor factorization. In: Proceedings of the 2010 SIAM International Conference on Data Mining, pp. 211–222. SIAM, Philadelphia (2010)
64. Yan, L., Li, W.J., Xue, G.R., Han, D.: Coupled group lasso for web-scale ctr prediction in display advertising. In: Proceedings of the 31th International Conference on Machine Learning, pp. 802–810. ACM, New York (2014)
65. Yang, Q., Fan, J., Wang, J., Zhou, L.: Personalizing web page recommendation via collaborative filtering and topic-aware markov model. In: ICDM, pp. 1145–1150. (2010)
66. Yang, S.H., Long, B., Smola, A., Sadagopan, N., Zheng, Z., Zha, H.: Like like alike: joint friendship and interest propagation in social networks. In: Proceedings of the 20th international conference on World Wide Web, pp. 537–546. International World Wide Web Conferences Steering Committee (2011)
67. Yu, F., Liu, Q., Wu, S., Wang, L., Tan, T.: A dynamic recurrent model for next basket recommendation. In: SIGIR, pp. 729–732. (2016)
68. Zhang, L., Agarwal, D., Chen, B.C.: Generalizing matrix factorization through flexible regression priors. In: Proceedings of the fifth ACM conference on Recommender systems, pp. 13–20. ACM, New York (2011)
69. Zhang, Y., Dai, H., Xu, C., Feng, J., Wang, T., Bian, J., Wang, B., Liu, T.Y.: Sequential click prediction for sponsored search with recurrent neural networks. In: AAAI, pp. 1369–1376. (2014)
70. Zheng, Y., Burke, R.D., Mobasher, B.: Splitting approaches for context-aware recommendation: an empirical study. In: Symposium on Applied Computing, SAC 2014, Gyeongju, Republic of Korea, pp. 274–279, 24–28 March, 2014
71. Zhong, E., Fan, W., Yang, Q.: Contextual collaborative filtering via hierarchical matrix factorization. In: Proceedings of the 2012 SIAM International Conference on Data Mining, pp. 744–755. SIAM, Philadelphia (2012)

Chapter 3
Contextual Operation

Abstract Motivated by recent works of natural language processing, this chapter introduces the concept of contextual operation for context-aware modeling. This operation represents each context value with a latent vector and models the contextual information as a semantic operation on the user and item. Besides, the contextual operating tensor is used to capture the common semantic effects of contexts. This chapter introduces notations and fundamental concepts of context representation, and then thoroughly presents the contextual operating tensor (COT) model. Finally, the process of parameter inference and the optimization algorithm is discussed.

3.1 Introduction

The state-of-the-art context modeling methods usually treat contexts as certain dimensions similar to those of users and items and capture relevance between contexts and users/items. Such kind of relevance sometimes is difficult to explain. Some works on multi-domain relation prediction can also be used for the context-aware collaborative prediction, but they have limitations in generating prediction under a large amount of contextual information.

To overcome the shortages of the existing methods mentioned above, we introduce a context modeling method contextual operating tensor model, which is motivated by the recent work of semantic compositionality in natural language processing (NLP). The context operating tensor (COT) method learns representation vectors of context values and uses contextual operations to capture the semantic operations of the contextual information. We provide a strategy for embedding each context value into a latent representation, no matter which domain the value belongs to. For each user-item interaction, we use contextual operating matrices to represent the semantic operations of these contexts and employ contextual operating tensors to capture common effects of contexts. Then, the operating matrix can be generated by multiplying latent representations of contexts with the operating tensor.

Parts of this chapter is reprinted from [1], with permission from IEEE.

© The Author(s) 2017
S. Wu et al., *Context-Aware Collaborative Prediction*, SpringerBriefs
in Computer Science, https://doi.org/10.1007/978-981-10-5373-3_3

We present contextual operating tensor (COT) model, for the context-aware collaborative prediction. First, we introduce notations and fundamental concepts of context representation, and then present COT thoroughly. Finally, we describe the process of parameter inference and the optimization algorithm.

3.2 Notations

In typical collaborative prediction systems, there is a user set U and an item set V. $\mathbf{u} \in \mathbb{R}^d$ and $\mathbf{v} \in \mathbb{R}^d$ are latent vectors of user u and item v, where d is the dimensionality. There are multiple contexts associated with users, items, and user-item interactions, such as *age, gender, occupation, releaseYear, director, genre, theater, time, companion*.

In this work, we divide these multiple contexts into user contexts $\mathscr{C}_1^U, \mathscr{C}_2^U, ...,$ item contexts $\mathscr{C}_1^V, \mathscr{C}_2^V, ...$ and interaction contexts $\mathscr{C}_1^I, \mathscr{C}_2^I, ...$ User contexts and item contexts indicate the attribute information associated with the user and item, while interaction contexts describe the situations of user-item interaction. For instance, in the scenario of a movie recommendation system, item contexts contain *title, length, releaseYear, director*, and *genre*, and interaction contexts contain *theater, time*, and *companion*, etc.

A specific context value c_m^i is a variable of context $\mathscr{C}_m^I$. Context values of user u, $c^u = \{c_1^u, c_2^u, ...\}$, are called user context combination, and context values of item v, $c^v = \{c_1^v, c_2^v, ...\}$, are named item context combination. The interaction contexts of a user-item rating are named interaction context combination $c^i = \{c_1^i, c_2^i, ...\}$. The rating that user u provides to item v under contextual information c can be written as $r_{u,v,c}$. The general contextual information c associated with rating $r_{u,v,c}$ is composed of user context combination c^u, item context combination c^v, and interaction context combination c^i.

The representation vector of a context value c_m^i is denoted as $\mathbf{h}_m^i \in \mathbb{R}^{d_c}$. Each context combination can be illustrated by a latent matrix which consists of latent vectors of context values. Then, user context combination c^u, item context combination c^v, and interaction context combination c^i can be represented as $\mathbf{H}^u = [\mathbf{h}_1^u, \mathbf{h}_2^u, ...] \in \mathbb{R}^{d_c \times |c^u|}$, $\mathbf{H}^v = [\mathbf{h}_1^v, \mathbf{h}_2^v, ...] \in \mathbb{R}^{d_c \times |c^v|}$, and $\mathbf{H}^i = [\mathbf{h}_1^i, \mathbf{h}_2^i, ...] \in \mathbb{R}^{d_c \times |c^i|}$, respectively, where $|c^u|$, $|c^v|$, and $|c^i|$ are the numbers of user contexts, item contexts, and interaction contexts.

3.3 Context Representation

There are various types of context values in practical collaborative prediction systems, such as categorical value, categorical set value, and numerical value. Here, we show how different types of context values can be transformed into corresponding latent representations.

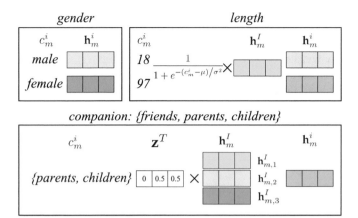

Fig. 3.1 Latent representations of context values in different domains. The context value in a categorical domain can be converted into a latent vector. Context values in one numerical domain have a shared latent vector and each value can be represented by a scalar multiplication. Each element in a categorical set is represented by a latent vector, and each context value can be represented by averaging latent vectors of elements in this context value

3.3.1 Categorical Domain

If a user watches a movie in a *theater*, *theater* is the categorical context value. Each context value c_m^i in the categorical domain should be represented by a distinct vector $\mathbf{h}_m^i \in \mathbb{R}^{d_c}$. Figure 3.1 shows *male* and *famale* in the categorical domain *gender* are represented by two distinct vectors.

3.3.2 Numerical Domain

Numerical context values widely exist, e.g., the *age* of a user, the *length* of a movie, and the *time* when a user watches a movie. To match with the representations of context values in other domains, we use a vector $\mathbf{h}_m^I \in \mathbb{R}^{d_c}$ to represent a numerical domain. To alleviate the dominant effect of large context values and the negligible effect of small ones, we employ a logistic function in normalization. Assume context values in a numerical domain falls a normal distribution, we can calculate the mean μ and variance σ^2 of this domain. Then a context value c_m^i in this domain is represented as a logistic function $\mathbf{h}_m^i = \left(1 + exp\left(-\left(c_m^i - \mu\right)/\sigma^2\right)\right)^{-1} \cdot \mathbf{h}_m^I$. For example, in Fig. 3.1, the numerical domain *length* is represented as $\mathbf{h}_m^I$, then 18 and 97 are normalized by a logistic function as $\left(1 + exp\left(-(18 - \mu)/\sigma^2\right)\right)^{-1} \cdot \mathbf{h}_m^I$ and $\left(1 + exp\left(-(97 - \mu)/\sigma^2\right)\right)^{-1} \cdot \mathbf{h}_m^I$, where μ and σ^2 are the corresponding mean and variance of the domain *length*.

3.3.3 Categorical Set Domain

When a user watches a movie with parents and children, this companion $c_m^i = \{parents, children\}$ is a context value in the categorical set domain $\mathscr{C}_m^I = \{friends, parents, children\}$. For this context value, we construct an indicator vector, where we normalize this vector for non-empty context values such that all values in the vector sum up to 1. Then, we estimate a latent vector $\mathbf{h}_{m,*}^i$ for each element $c_{m,*}^i$ in $\mathscr{C}_m^I$. For example, in Fig. 3.1, watching a movie with *parents* and *children*, the indicator vector is $\mathbf{z}^T = (0, 0.5, 0.5)$ and the categorical set domain $\{friends, parents, children\}$ is represented by $(\mathbf{h}_{m,1}^I; \mathbf{h}_{m,2}^I; \mathbf{h}_{m,3}^I)$. Then, the context value $\{parents, children\}$ can be computed as $\mathbf{h}_m^i = \mathbf{z}^T \cdot (\mathbf{h}_{m,1}^I; \mathbf{h}_{m,2}^I; \mathbf{h}_{m,3}^I)$.

In practical applications, various types of contexts fall into one of these domains mentioned above. For instance, the geographical location can be denoted by a numerical set $\{latitude, longitude\}$, where each element has its numerical domain. Some kinds of entity contexts, for instance, low-level features of image/text and social relations, can be transformed into feature vectors using machine learning techniques. Each value in the obtained feature vector can be treated as the value in a numerical domain.

3.4 Contextual Operating Matrix

In typical matrix factorization methods, latent vectors of users and items are constant with varying contexts. But in real-world applications, user interests, and item properties are changed with varying contexts. Here, under different contexts, we provide context-specific latent vectors for users and items, and the rating prediction can be rewritten as:

$$\hat{r}_{u,v,c} = b_0 + b_u + b_v + \sum_{m=1}^{|c|} b_{c,m} + \mathbf{u}_c^T \mathbf{v}_c , \tag{3.1}$$

where $|c| = |c^u| + |c^v| + |c^i|$ is the number of contexts, b_0 is the mean rating in training data, b_u and b_v denote the biases of user u and item v, $b_{c,m}$ is the bias of a context value. $\mathbf{u}_c$ and $\mathbf{v}_c$ are latent vectors of user u and item v under the contextual information c.

In a phrase of noun and adjective, the noun has semantic information and the adjective has semantic operation on the noun. In collaborative prediction systems, entities have rich semantic information and contexts act like adjectives which have the semantic operation on entities. For example, companion with children can change interests of a user, and he/she may tend to watch cartoons with children. During Valentine's Day, the latent characteristics of a romantic film may be changed and this film may become popular. We use contextual operating matrices to reveal how the contextual information c affects the properties of user/item. The context-specific latent vectors of users and items can be generated from their original ones.

$$\mathbf{u}_c = \mathbf{M}_c^U \mathbf{u} , \tag{3.2}$$

$$\mathbf{v}_c = \mathbf{M}_c^V \mathbf{v} , \tag{3.3}$$

where $\mathbf{u}$ and $\mathbf{v}$ are the original vectors of the user and item, $\mathbf{M}_c^U$ and $\mathbf{M}_c^V$ are $d \times d$ contextual operating matrices of the contextual information c on users and items. These context-specific vectors can also be generated from other nonlinear function, e.g., the sigmoid function in Eqs. 3.4 and 3.5, to assess the effectiveness of the COT framework. In the experimental section, we will show the results of these two computations.

$$\mathbf{u}_c = \left(1 + exp\left(-\mathbf{M}_c^U \mathbf{u}\right)\right)^{-1} - 0.5 , \tag{3.4}$$

$$\mathbf{v}_c = \left(1 + exp\left(-\mathbf{M}_c^V \mathbf{v}\right)\right)^{-1} - 0.5 , \tag{3.5}$$

Comparing with the interaction context combination which can operate the latent properties of users and items simultaneously, the user or item context combination does not have very similar operation on both users and items. For instance, a user context *occupation* can indicate the potential characteristics of users which may not have been revealed by the observed user-item interactions, but has a slight influence in changing item properties. In this work, confronting with three kinds of contextual information, we plan to separate their effects.

3.4.1 Linear Computation

Treating three context combinations separately, we can rewrite the contextual operating matrix as an operation combination. Serial multiplication may enlarge the defect of one context combination. Here, we resort to the linear computation, contextual operating matrices of users and items are denoted as

$$\mathbf{M}_c^U = \mathbf{M}_{c_u}^U + \mathbf{M}_{c_v}^U + \mathbf{M}_{c_i}^U, \tag{3.6}$$

$$\mathbf{M}_c^V = \mathbf{M}_{c_u}^V + \mathbf{M}_{c_v}^V + \mathbf{M}_{c_i}^V, \tag{3.7}$$

where $\mathbf{M}_{c_u}^U$, $\mathbf{M}_{c_v}^U$, and $\mathbf{M}_{c_i}^U$ are user-wise operation matrices of user context combination c^u, item context combination c^v, and interaction context combination c^i. $\mathbf{M}_{c_u}^V$, $\mathbf{M}_{c_v}^V$ and $\mathbf{M}_{c_i}^V$ are item-wise operation matrices of these context combinations.

3.4.2 Nonlinear Computation

This linear computation not only can separate the effect of three context combinations but also can learn different weights of them implicitly. It can be replaced by other nonlinear computation. For example, using a sigmoid function, these two operation matrices can be denoted as

$$\mathbf{M}_c^U = \left(1 + exp\left(-\mathbf{M}_{c_u}^U - \mathbf{M}_{c_v}^U - \mathbf{M}_{c_i}^U\right)\right)^{-1} - 0.5 \,, \qquad (3.8)$$

$$\mathbf{M}_c^V = \left(1 + exp\left(-\mathbf{M}_{c_u}^V - \mathbf{M}_{c_v}^V - \mathbf{M}_{c_i}^V\right)\right)^{-1} - 0.5, \qquad (3.9)$$

3.5 Contextual Operating Tensor

We need two weighting matrices to map the latent matrix of a specific context combination into the operation matrices. For example, we need to estimate two matrices for each $\mathbf{H}^i$ and obtain operation matrices $\mathbf{M}_{c_i}^U$ and $\mathbf{M}_{c_i}^V$. The number of parameters will increase rapidly as the number of context combinations grows. Besides, since different contexts share similar semantic effects, for example, both *weekend* and being at *home* may make you would like to read novels. It will be plausible if we can generate contextual operating matrices from several basic matrices (operating tensor) which represent some common semantic effects of contexts.

To employ the contextual operating tensor, we first should convert the latent matrix of context combination into a vector, and then the operating matrix can be generated from the multiplication of this vector with the operating tensor. Here, we show how to transform latent matrices of three context combinations into latent vectors as follows:

$$\mathbf{a}_{c_u}^U = \mathbf{H}^u \mathbf{w}_{C_U}^U, \ \mathbf{a}_{c_v}^U = \mathbf{H}^v \mathbf{w}_{C_V}^U, \ \mathbf{a}_{c_i}^U = \mathbf{H}^i \mathbf{w}_{C_I}^U,$$

$$\mathbf{a}_{c_u}^V = \mathbf{H}^u \mathbf{w}_{C_U}^V, \ \mathbf{a}_{c_v}^V = \mathbf{H}^v \mathbf{w}_{C_V}^V, \ \mathbf{a}_{c_i}^V = \mathbf{H}^i \mathbf{w}_{C_I}^V,$$

where each column of $\mathbf{H}^i$ denotes the latent vector of a context value, and $\mathbf{w}_{C_I}^U$ indicates the user-wise context weights on $\mathbf{H}^i$. The context combination vector $\mathbf{a}$ is a d_c dimensional latent vector, which is a weighted combination of context vectors.

We demonstrate the process of mapping the latent matrix of a context combination to vectors in Fig. 3.2. For latent matrix $\mathbf{H}^i$, we estimate two weighting vectors $\mathbf{w}_{C_I}^U$ and $\mathbf{w}_{C_I}^V$ which indicate user-wise and item-wise weights on this context combination. Multiplying the latent matrix with the weighting vectors, we can obtain the user-specific and item-specific latent vectors $\mathbf{a}_{C_I}^U$ and $\mathbf{a}_{C_I}^V$, respectively. For example, given *Tom* and *Titanic*, the contextual information of *theater*, *time*, *weather*, *companion* is shown in the middle part, and each context is denoted as a column. Context weights

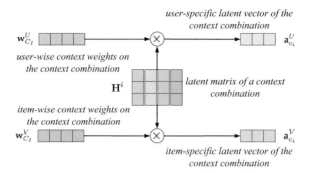

Fig. 3.2 The process of generating latent vectors of a context combination. The middle part is the latent matrix $\mathbf{H}^i$ of the context combination c^i. The left part is user-wise and item-wise context weights which indicate different influences of context values on users and items. The right part denotes the user-specific and item-specific latent vectors of this context combination

in the left indicate the influences of four contexts on *Tom* and *Titanic*, the right part is context combination vectors for *Tom* and *Titanic*.

After obtaining latent vectors of context combinations, operation matrices can be generated by multiplying the latent vectors of context combinations with the operating tensors. We use $\mathbf{T}^{U,[1:d]}$ and $\mathbf{T}^{V,[1:d]}$ to denote the operating tensors for users and items, and briefly write as $\mathbf{T}^U$ and $\mathbf{T}^V$ for simplicity. Since entity contexts have significantly different properties from interaction contexts, we would like to employ different operating tensors for three context combinations. The contextual operation matrices are calculated as

$$\mathbf{M}_c^U = \left(\mathbf{a}_{c_u}^U\right)^T \mathbf{T}_{C_U}^U + \left(\mathbf{a}_{c_v}^U\right)^T \mathbf{T}_{C_V}^U + \left(\mathbf{a}_{c_i}^U\right)^T \mathbf{T}_{C_I}^U , \qquad (3.10)$$

$$\mathbf{M}_c^V = \left(\mathbf{a}_{c_u}^V\right)^T \mathbf{T}_{C_U}^V + \left(\mathbf{a}_{c_v}^V\right)^T \mathbf{T}_{C_V}^V + \left(\mathbf{a}_{c_i}^V\right)^T \mathbf{T}_{C_I}^V , \qquad (3.11)$$

where $\mathbf{T}_{C_U}^U$, $\mathbf{T}_{C_V}^U$, and $\mathbf{T}_{C_I}^U$ are $d_c \times d \times d$ tensors, denoting the operating tensors of three context combinations for users, $\mathbf{T}_{C_U}^V$, $\mathbf{T}_{C_V}^V$, and $\mathbf{T}_{C_I}^V$ are operating tensors for items. The operating tensor is composed of d slices, and each slice should be multiplied with the vector of context combination. Substituting Eqs. (3.10)–(3.11) in Eqs. (3.2)–(3.3), we write detailed equations of context-specific latent vectors of users and items.

$$\mathbf{u}_c = \begin{bmatrix} \left(\mathbf{a}_{c_u}^U\right)^T \mathbf{T}_{C_U,1}^U \mathbf{u} + \left(\mathbf{a}_{c_v}^U\right)^T \mathbf{T}_{C_V,1}^U \mathbf{u} + \left(\mathbf{a}_{c_i}^U\right)^T \mathbf{T}_{C_I,1}^U \mathbf{u} \\ \cdots \\ \left(\mathbf{a}_{c_u}^U\right)^T \mathbf{T}_{C_U,d}^U \mathbf{u} + \left(\mathbf{a}_{c_v}^U\right)^T \mathbf{T}_{C_V,d}^U \mathbf{u} + \left(\mathbf{a}_{c_i}^U\right)^T \mathbf{T}_{C_I,d}^U \mathbf{u} \end{bmatrix}$$

$$\mathbf{v}_c = \begin{bmatrix} \left(\mathbf{a}_{c_u}^V\right)^T \mathbf{T}_{C_U,1}^V \mathbf{v} + \left(\mathbf{a}_{c_v}^V\right)^T \mathbf{T}_{C_V,1}^V \mathbf{v} + \left(\mathbf{a}_{c_i}^V\right)^T \mathbf{T}_{C_I,1}^V \mathbf{v} \\ \cdots \\ \left(\mathbf{a}_{c_u}^V\right)^T \mathbf{T}_{C_U,d}^V \mathbf{v} + \left(\mathbf{a}_{c_v}^V\right)^T \mathbf{T}_{C_V,d}^V \mathbf{v} + \left(\mathbf{a}_{c_i}^V\right)^T \mathbf{T}_{C_I,d}^V \mathbf{v} \end{bmatrix}$$

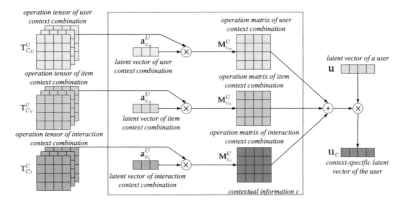

Fig. 3.3 Overview of constructing the context-specific latent vector for a user. Contextual operating tensors are shown on the left side, and the process of generating the operating matrix is illustrated in the square

where $\mathbf{T}^U_{C_I,m}$ and $\mathbf{T}^V_{C_I,m}$ are $d_c \times d$ matrices, denoting the mth slice of $\mathbf{T}^U_{C_I}$ and $\mathbf{T}^V_{C_I}$. Each slice captures a specific type of common semantic operation on users/items.

The process of generating the context-specific latent vector of a user is illustrated in Fig. 3.3. After generating the latent vector of contexts in Fig. 3.2, we can obtain the operation of these contexts by multiplying with a tensor. The operation matrix of these contexts should change the latent vector of user or item under these contexts. There are three operating tensors $\mathbf{T}^U_{C_U}$, $\mathbf{T}^U_{C_V}$, and $\mathbf{T}^U_{C_I}$. For a specific user-item interaction $r_{u,v,c}$, we use $\mathbf{a}^U_{c_u}$, $\mathbf{a}^U_{c_v}$ and $\mathbf{a}^U_{c_i}$ to represent three context combinations. Multiplying these vectors of context combinations with the tensors, the corresponding operation matrices $\mathbf{M}^U_{c_u}$, $\mathbf{M}^U_{c_v}$, and $\mathbf{M}^U_{c_i}$ can be obtained, which are certain combinations of semantic operations in respective contextual operating tensors. The linear computation of three operation matrices $\mathbf{M}^U_{c_u} + \mathbf{M}^U_{c_v} + \mathbf{M}^U_{c_i}$ is used to change the original latent vector of user u. For instance, given *Tom*, *Titanic* and contexts *theater*, *time*, *weather*, *companion*, we have shown how to generate context combination vectors for *Tom* and *Titanic* in Fig. 3.3. In Fig. 3.3, we can compute the operation matrix of these vectors by multiplying with operating tensor in the left. Then the latent vector of *Tom* under these contexts can be calculated by using operation matrix and the original latent vector.

After discussing the generating process of context-specific vectors of users and items, the overall prediction function of COT can be written as:

$$\hat{r}_{u,v,c} = b_0 + b_u + b_v + \sum_{m=1}^{|c|} b_{c,m} +$$
$$\left\{ \left[\left(\mathbf{a}^U_{c_u}\right)^T \mathbf{T}^U_{C_U} + \left(\mathbf{a}^U_{c_v}\right)^T \mathbf{T}^U_{C_V} + \left(\mathbf{a}^U_{c_i}\right)^T \mathbf{T}^U_{C_I} \right] \mathbf{u} \right\}^T$$
$$\left\{ \left[\left(\mathbf{a}^V_{c_u}\right)^T \mathbf{T}^V_{C_U} + \left(\mathbf{a}^V_{c_v}\right)^T \mathbf{T}^V_{C_V} + \left(\mathbf{a}^V_{c_i}\right)^T \mathbf{T}^V_{C_I} \right] \mathbf{v} \right\}$$

3.5.1 Parameter Inference

We have already introduced our model mathematically in the previous section. Now, to accomplish the parameter inference, we need to minimize the following objective function:

$$
\min_{\mathbf{u},\mathbf{v},\mathbf{H},\mathbf{T},\mathbf{w}} J = \sum_{\langle u,v,c\rangle \in \Omega} (r_{u,v,c} - \hat{r}_{u,v,c})^2
$$

$$
+ \frac{\lambda}{2}(b_u{}^2 + b_v{}^2 + \sum_{m=1}^{|c|} b_{c,m}{}^2 \tag{3.12}
$$

$$
+ ||\mathbf{u}||^2 + ||\mathbf{v}||^2 + ||\mathbf{H}||^2 + ||\mathbf{T}||^2 + ||\mathbf{w}||^2),
$$

where Ω denotes the training set, and λ is a parameter to control the regularizations, which can be determined using cross-validation. The derivations of J with respect to all parameters can be calculated as:

$$
\frac{\partial J}{\partial b_*} = -2l_{u,v,c} + \lambda b_* ,
$$

$$
\frac{\partial J}{\partial \mathbf{u}} = -2l_{u,v,c} \left(\mathbf{M}_c^U\right)^T \left(\mathbf{M}_c^V \mathbf{v}\right) + \lambda \mathbf{u} ,
$$

$$
\frac{\partial J}{\partial \mathbf{v}} = -2l_{u,v,c} \left(\mathbf{M}_c^U \mathbf{u}\right) \mathbf{M}_c^V + \lambda \mathbf{v} ,
$$

$$
\frac{\partial J}{\partial \mathbf{H}_*} = -2l_{u,v,c} \left(\mathbf{T}_{C_*}^U \mathbf{u}\right) \left(\mathbf{M}_c^V \mathbf{v}\right) \left(\mathbf{w}_{C_*}^U\right)^T
$$

$$
+ l_{u,v,c} \left(\mathbf{T}_{C_*}^V \mathbf{v}\right) \left(\mathbf{M}_c^U \mathbf{u}\right) \left(\mathbf{w}_{C_*}^V\right)^T + \lambda \mathbf{H}_* ,
$$

$$
\frac{\partial J}{\partial \mathbf{w}_{C_U}^U} = -2l_{u,v,c} \mathbf{H}_u^T \left(\mathbf{T}_{C_U}^U \mathbf{u}\right) \left(\mathbf{M}_c^V \mathbf{v}\right) + \lambda \mathbf{w}_{C_U}^U ,
$$

$$
\frac{\partial J}{\partial \mathbf{w}_{C_V}^U} = -2l_{u,v,c} \mathbf{H}_v^T \left(\mathbf{T}_{C_V}^U \mathbf{u}\right) \left(\mathbf{M}_c^V \mathbf{v}\right) + \lambda \mathbf{w}_{C_V}^U ,
$$

$$
\frac{\partial J}{\partial \mathbf{w}_{C_I}^U} = -2l_{u,v,c} \mathbf{H}_i^T \left(\mathbf{T}_{C_I}^U \mathbf{u}\right) \left(\mathbf{M}_c^V \mathbf{v}\right) + \lambda \mathbf{w}_{C_I}^U ,
$$

$$
\frac{\partial J}{\partial \mathbf{w}_{C_U}^V} = -2l_{u,v,c} \left(\mathbf{M}_c^U \mathbf{u}\right) \mathbf{H}_u^T \left(\mathbf{T}_{C_U}^V \mathbf{u}\right) + \lambda \mathbf{w}_{C_U}^V ,
$$

$$
\frac{\partial J}{\partial \mathbf{w}_{C_V}^V} = -2l_{u,v,c} \left(\mathbf{M}_c^U \mathbf{u}\right) \mathbf{H}_v^T \left(\mathbf{T}_{C_V}^V \mathbf{u}\right) + \lambda \mathbf{w}_{C_V}^V ,
$$

$$\frac{\partial J}{\partial \mathbf{w}_{C_l}^V} = -2l_{u,v,c}\left(\mathbf{M}_c^U \mathbf{u}\right)\mathbf{H}_i^T\left(\mathbf{T}_{C_l}^V \mathbf{u}\right) + \lambda \mathbf{w}_{C_l}^V \,,$$

$$\frac{\partial J}{\partial \mathbf{T}_{*,m}^U} = -2l_{i,j,k}\mathbf{H}_u \mathbf{w}_*^U \mathbf{u}^T v_{c,m} + \lambda \mathbf{T}_{*,m}^U \,,$$

$$\frac{\partial J}{\partial \mathbf{T}_{*,m}^V} = -2l_{i,j,k}\mathbf{H}_v \mathbf{w}_*^V \mathbf{v}^T u_{c,m} + \lambda \mathbf{T}_{*,m}^V \,,$$

where b_* is a specific bias, $\mathbf{H}_*$ describes the latent matrix of a specific context combination, $\mathbf{T}_*^U$ is an operating tensor of a specific context combination for the user, and $\mathbf{T}_{*,m}^U$ is the mth slide of the operating tensor $\mathbf{T}_*^U$. $u_{c,m}$ and $v_{c,m}$ denote the mth component of latent vector $\mathbf{u}_c$ and $\mathbf{v}_c$, respectively, and $l_{u,v,c} = r_{u,v,c} - \hat{r}_{u,v,c}$.

3.5.2 Optimization Algorithms

After calculating all the derivations, a minimum solution of J in Eq. 3.12 can be obtained by using stochastic gradient descent, which has been widely used in recommender systems [2, 3]. We propose an efficient learning algorithm (Algorithm 1) to optimize the objective function with the contextual operation. At first, all the parameters are initialized randomly in the range $[-0.5, 0.5]$. Then, we randomly choose a rating $r_{u,v,c}$ from the training set and update all parameters using the derivations in the section of parameter inference. After the algorithm is convergent, the model parameters b, $\mathbf{u}$, $\mathbf{v}$, $\mathbf{H}$, $\mathbf{T}$, and $\mathbf{w}$ are obtained, and the rating prediction $\hat{r}_{u,v,c}$ can be calculated using Eq. 5.13. Note that γ is the learning rate, which can be determined through the cross-validation. This optimization algorithm can be implemented without requiring significant change to conventional matrix factorization models.

3.5.3 Complexity Analysis

Based on the optimization algorithm, now we analyze the time complexity of training process. In each iteration, the time complexity of updating $\mathbf{u}$ and $\mathbf{v}$ are $O(d^2 \times |\Omega|)$, where $|\Omega|$ is the size of training dataset. The time complexity of updating $\mathbf{H}$ and $\mathbf{w}$ are $O(d_c \times d^2 \times |\Omega|)$, and the complexity of updating $\mathbf{T}$ is $O(d_c \times d \times |\Omega|)$. Therefore, the total time complexity of training process is $O(d_c \times d^2 \times |\Omega|)$. Since $|\Omega|$ is much larger than $d_c \times d^2$, the time complexity can be viewed as growing linearly with respect to the size of training dataset. Therefore, the time complexity of COT is very similar to that of the state-of-the-art CARS[2] and FM models, which both can be treated as linear with size of training set. This time complexity also shows that COT has potential to scale up to large-scale data sets.

Algorithm 1 Optimization Algorithm of COT

1: **Input**: The training set, each $r_{u,v,c}$ is associated with a user u, an item v and contextual information c.
2: **Output**: Model parameters b, **u**, **v**, **H**, **T** and **w**.
3: Initialize b, **u**, **v**, **H**, **T** and **w** randomly.
4: **while** not convergent **do**
5:    Select an instance $r_{u,v,c}$ from the training set.
6:    Calculate $\frac{\partial J}{\partial b}$, $\frac{\partial J}{\partial \mathbf{u}}$, $\frac{\partial J}{\partial \mathbf{v}}$, $\frac{\partial J}{\partial \mathbf{H}}$, $\frac{\partial J}{\partial \mathbf{T}}$, $\frac{\partial J}{\partial \mathbf{w}}$.
7:    Update $b \leftarrow b - \gamma \frac{\partial J}{\partial b}$.
8:    Update $\mathbf{u} \leftarrow \mathbf{u} - \gamma \frac{\partial J}{\partial \mathbf{u}}$.
9:    Update $\mathbf{v} \leftarrow \mathbf{v} - \gamma \frac{\partial J}{\partial \mathbf{v}}$.
10:   Update $\mathbf{H} \leftarrow \mathbf{H} - \gamma \frac{\partial J}{\partial \mathbf{H}}$.
11:   Update $\mathbf{T} \leftarrow \mathbf{T} - \gamma \frac{\partial J}{\partial \mathbf{T}}$.
12:   Update $\mathbf{w} \leftarrow \mathbf{w} - \gamma \frac{\partial J}{\partial \mathbf{w}}$.
13: **end while**

3.6 Conclusion

This chapter presents a novel context-aware method, i.e., COT. This method provides each context value with a continuous vector, which is a distributed representation different from the one hot representation in FM and other methods. Such distributed representations have a powerful ability in describing the semantic operation of context values. Similar to the semantic composition in NLP where the adjective has an operation on the noun, we provide the contextual information of each rating event with a semantic operation matrix, which can be used to generate new vectors of users and items under this contextual situation. In addition, the contextual operating tensor is used to capture the common semantic effects of contexts. The contextual operating matrix can be calculated from the contextual operating tensor and context representations.

References

1. Wu, S., Liu, Q., Wang, L., Tan, T.: Contextual operation for recommender systems. IEEE TKDE **28**, 2000–2012 (2016)
2. Koren, Y., Bell, R.: Advances in collaborative filtering. In: Recommender Systems Handbook, pp. 145–186. Springer, Berlin (2011)
3. Koren, Y., Bell, R., Volinsky, C.: Matrix factorization techniques for recommender systems. Computer **42**(8), 30–37 (2009)

Chapter 4
Hierarchical Representation

Abstract This chapter introduces a hierarchical interaction representation (HIR) model, which treats the interaction among different entities and contexts as representation. This model generates the interaction representation of two entities via tensor multiplication, which is performed iteratively to construct a hierarchical structure among all entities and contexts. Moreover, the model employs several hidden layers to reveal the underlying properties of this interaction representation and enhance the model performance further. After generating the final representation, the prediction can be calculated using a variety of machine learning methods according to different application tasks (e.g., linear regression for regression tasks, pair-wise ranking method for ranking tasks, and logistic regression for classification tasks).

4.1 Introduction

Matrix factorization [2, 3] is widely used for collaborative prediction between users and items. And matrix factorization-based methods have been extensively studied, some of which, e.g., factorization machine and tensor factorization, are implemented for predicting the multiple entities and contexts interaction. Both tensor factorization and factorization machine predict the interaction based on the similarity among all the entities and contexts, which seems intuitive and reasonable. However, this does not conform to situations of various applications. Using latent collaborative retrieval as an example, both tensor factorization and factorization machine compute the similarity among user, query, and document to predict their interaction. The similarity between the user and the query does not contribute to the document selection of this user, and the final selection is not merely based on the similarity between the user and the document and the similarity between the query and the document. We should measure the correlation between the document and the joint representation of user and query.

There is another disadvantage of conventional methods. They only calculate the similarity based on corresponding dimensional values in the latent vectors, which

Parts of this chapter is reprinted from [1], with permission from ACM.

S. Wu et al., *Context-Aware Collaborative Prediction*, SpringerBriefs
in Computer Science, https://doi.org/10.1007/978-981-10-5373-3_4

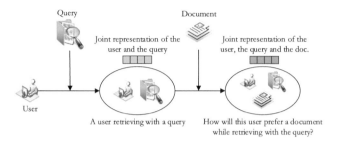

Fig. 4.1 Illustration of the hierarchical structure of multi-entity interaction, using latent collaborative retrieval as an example

does not allow rich high-order calculation among values of different dimensions. Therefore, to be general, we provide a latent representation for the interaction of entities and contexts, which describes how these entities and contexts would act being together with each other and can reveal their high-order relation. As illustrated in Fig. 4.1, when a user retrieves a query, his or her property will be changed and form a joint representation of the user and the query. Similarly, when the user views a document, there will be a joint representation of user, query, and document, which indicates how will this user prefers a document while retrieving with the query. Based on this joint representation, we could make prediction for the interaction of entities and contexts.

We present the hierarchical interaction representation (HIR) model for predicting interaction among multiple entities and contexts. Each entity and context is represented as a latent vector. Three-dimensional tensor multiplication captures the joint characteristics of entities and contexts.

In this chapter, we introduce the hierarchical interaction representation. We first introduce the definition and notations of the problem at first. Then, we present the interaction representation of the entity and context, followed by the hierarchical representation of more entities and contexts. Finally, we show how to enhance the model performance via employing hidden layers.

4.2 Notations

The problem we study here can be described as follows. In an interaction scenario, suppose that we have a collaborative prediction task with n types of entities and contexts denoted by $\{E^{(1)}, ..., E^{(n)}\}$. For each type of entity or context, we have $E^{(m)} = \{e_1^{(m)}, e_2^{(m)}, ...\}$, $e_i^{(m)} \in \mathbb{R}^d$, $m \in \{1, ..., n\}$. In the interaction of entities and contexts, the relation among $e_{k_1}^{(1)}, ..., e_{k_n}^{(n)}$ is denoted by $y_{k_1,...,k_n}$. The task of collaborative prediction is to give a prediction $\hat{y}_{k_1,...,k_n}$ based on all the entities and contexts $e_{k_1}^{(1)}, ..., e_{k_n}^{(n)}$.

Here, using latent collaborative retrieval as an example, there will be three entities and contexts (i.e., user, query, document) denoted by $\{E^{(1)}, E^{(2)}, E^{(3)}\}$. Specifically, users are denoted by $E^{(1)} = \{e_1^{(1)}, e_2^{(1)}, ...\}$, and each user is denoted as $e_i^{(1)} \in \mathbb{R}^d$. Queries are denoted by $E^{(2)} = \{e_1^{(2)}, e_2^{(2)}, ...\}$, and each query is denoted as $e_i^{(2)} \in \mathbb{R}^d$. Documents are denoted by $E^{(3)} = \{e_1^{(3)}, e_2^{(3)}, ...\}$, and each document is denoted as $e_i^{(3)} \in \mathbb{R}^d$. Then the interaction among user, query and document is denoted as y_{k_1,k_2,k_3}.

In this work, we would jointly model all the entities and contexts using an interaction representation $r_{k_1,...,k_n}^{(n)}$, and m-subset of the entities and contexts with a interaction representation $r_{k_1,...,k_m}^{(m)}$. The interaction representation describes the joint characteristics of the entities and contexts. We could make prediction of their interaction based on the interaction representation using various learning methods (i.e., linear regression, logistic regression, and pair-wise ranking method).

4.3 Representation of Entities and Contexts

4.3.1 Interaction Representation

We first introduce the situation with two entities (contexts, or an entity and a context). To get the joint representation of two entities, we need to employ an interaction function that satisfies:

$$r_{k_1,k_2}^{(2)} = f^{(2)}\left(e_{k_1}^{(1)}, e_{k_2}^{(2)}\right) , \tag{4.1}$$

where $f^{(2)}(\cdot)$ denotes an interaction function of two entities. $r_{k_1,k_2}^{(2)}$ describes how the two entities would behave being together in an interaction scenario. For instance, if there are user and movie as two entities, their joint representation captures the characteristic of a user watching a movie, and models how much he or she likes the movie.

To capture the high-order relation between entities, we use tensor multiplication, as shown in the bottom square of Fig. 4.2. Here, we compute the interaction representation as:

$$r_{k_1,k_2}^{(2)} = [\left(e_{k_2}^{(2)}\right)^T T^{(1)}]e_{k_1}^{(1)} , \tag{4.2}$$

where $T^{(1)}$ is a $d \times d \times d$ dimensional interaction tensor, modeling the interaction of these two entities. The tensor multiplication $\left(e_{k_2}^{(2)}\right)^T T^{(1)}$ can be defined as:

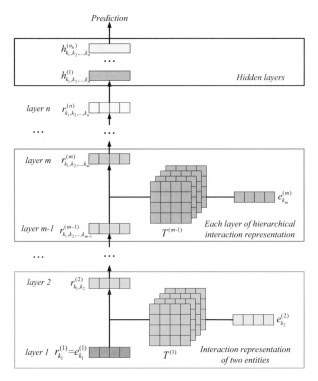

Fig. 4.2 Overview of the learning procedure of HIR. Interaction representation of two entities is shown in the bottom square. Each layer of HIR is shown in the middle square. The top part illustrates hidden layers between the interaction representation and the prediction

$$
\left(e_{k_2}^{(2)} \right)^T T^{(1)} =
\begin{bmatrix}
\left(e_{k_2}^{(2)} \right)^T T_1^{(1)} \\
\cdots \\
\left(e_{k_2}^{(2)} \right)^T T_d^{(1)}
\end{bmatrix},
\tag{4.3}
$$

and Eq. (4.2) can be calculated as:

$$
r_{k_1,k_2}^{(2)} =
\begin{bmatrix}
\left(e_{k_2}^{(2)} \right)^T T_1^{(1)} e_{k_1}^{(1)} \\
\cdots \\
\left(e_{k_2}^{(2)} \right)^T T_d^{(1)} e_{k_1}^{(1)}
\end{bmatrix},
\tag{4.4}
$$

where $T_m^{(1)}$ denotes the mth slice of the tensor T and each slice is a $d \times d$ dimensional matrix.

With such computation, the interaction tensor can model the rich interaction between entities. Each slice of the interaction tensor can capture a specific type of interaction between these entities.

Now, to predict the interaction of the two entities, we can use a simple linear regression to generate the prediction, which can be concluded as:

$$\hat{y}_{k_1,k_2} = W^T [(e_{k_2}^{(2)})^T T^{(1)}] e_{k_1}^{(1)} , \qquad (4.5)$$

where W is a n dimensional vector denoting the weights of the regression.

4.3.2   Hierarchical Interaction Representation

After introducing the joint representation of two entities, we would model the joint representation of more entities and contexts in general situation. For more entities and contexts, we need another interaction function that satisfies:

$$r_{k_1,...,k_n}^{(n)} = f^{(n)} \left(e_{k_1}^{(1)}, ..., e_{k_n}^{(n)} \right) , \qquad (4.6)$$

where $f^{(n)}(\cdot)$ denotes an interaction function of n entities and contexts. $r_{k_1,...,k_n}^{(n)}$ describes the joint characteristics of all the entities and contexts in an interaction scenario.

To generate the interaction representation of n entities and contexts, we perform the tensor multiplication recursively, which forms a hierarchical structure with n layers. These procedures can be concluded as:

$$\begin{cases} r_{k_1}^{(1)} = e_{k_1}^{(1)} \\ r_{k_1,...,k_m}^{(m)} = [(e_{k_m}^{(m)})^T T^{(m-1)}] r_{k_1,...,k_{m-1}}^{(m-1)} \end{cases} , \qquad (4.7)$$

where $1 < m \leq n$, and $T^{(m-1)}$ is a $d \times d \times d$ dimensional interaction tensor, generating the interaction representation in the mth layer. Each layer of HIR is illustrated in the middle square of Fig. 4.2. Thus, the joint representation of all entities and contexts can be calculated as:

$$r_{k_1,...,k_n}^{(n)} = \prod_{i=2}^{n} [(e_{k_i}^{(i)})^T T^{(i-1)}] e_{k_1}^{(1)} . \qquad (4.8)$$

Using latent collaborative retrieval as an example, there are three entities and contexts, i.e., user, query, and document. Their interaction representation captures how a user prefers a document while retrieving with a query, which can be calculated as:

$$r^{(3)}_{k_1,k_2,k_3} = [\left(e^{(3)}_{k_3}\right)^T T^{(2)}][\left(e^{(2)}_{k_2}\right)^T T^{(1)}]e^{(1)}_{k_1} . \qquad (4.9)$$

With the joint representation, we can compute the prediction of the multiply entities and contexts. A variety of machine learning methods can be used according to the specific task. For example, with a linear regression method, the prediction can be generated as:

$$\hat{y}_{k_1,\dots,k_n} = W^T \prod_{i=2}^{n} [\left(e^{(i)}_{k_i}\right)^T T^{(i-1)}]e^{(1)}_{k_1} , \qquad (4.10)$$

where W is a n-dimensional vector denoting the weights of the regression. Moreover, as another example, in tasks with three entities or contexts, the prediction can be generated as:

$$\hat{y}_{k_1,k_2,k_3} = W^T [\left(e^{(3)}_{k_3}\right)^T T^{(2)}][\left(e^{(2)}_{k_2}\right)^T T^{(1)}]e^{(1)}_{k_1} . \qquad (4.11)$$

4.4   Multiple Hidden Layers

Deep models with multiple layers have shown delightful performance in different areas [4]. As shown in the top square of Fig. 4.2, to enhance the ability of representation and to improve the performance of our model, we can add several hidden layers between the joint representation $r^{(n)}_{k_1,\dots,k_n}$ and the prediction $\hat{y}_{k_1,\dots,k_n}$.

Suppose we add n_h hidden layers between the final representation and the prediction. With the effect of each hidden layer captured by a matrix, the joint representation of entities and contexts after hidden layers can be calculated iteratively as:

$$\begin{cases} h^{(0)}_{k_1,k_2,\dots,k_n} = r^{(n)}_{k_1,k_2,\dots,k_n} \\ h^{(m_h)}_{k_1,k_2,\dots,k_n} = H_{m_h} h^{(m_h-1)}_{k_1,k_2,\dots,k_n} \end{cases} , \qquad (4.12)$$

where $1 \leq m_h \leq n_h$, $h^{(m_h)}_{k_1,k_2,\dots,k_n}$ denotes the joint representation of entities and contexts after m_h hidden layers, and H_{m_h} is a $d \times d$ dimensional matrix of the m_hth hidden layer. If we write the whole procedure in one equation, the overall formulation generating the joint representation becomes:

$$h^{(n_h)}_{k_1,\dots,k_n} = \prod_{i=1}^{n_h} H_i \prod_{i=2}^{n} [\left(e^{(i)}_{k_i}\right)^T T^{(i-1)}]e^{(1)}_{k_1} . \qquad (4.13)$$

Furthermore, with a linear regression method, the final prediction can be made via:

$$\hat{y}_{k_1,\dots,k_n} = W^T \prod_{i=1}^{n_h} H_i \prod_{i=2}^{n} [\left(e^{(i)}_{k_i}\right)^T T^{(i-1)}]e^{(1)}_{k_1} , \qquad (4.14)$$

We implement HIR with 0, 1, and 2 hidden layers denoted as HIR, HIR+, and HIR++, respectively.

4.5   Learning for Context-Aware Prediction

In this section, we introduce the learning process of hierarchical interaction representation model in three tasks, i.e., regression with explicit feedback, ranking with implicit feedback, and classification for probability prediction.

4.5.1   Regression Task

In scenarios with explicit feedbacks, such as the rating prediction, the collaborative prediction can be calculated via a linear regression:

$$\hat{y}_{k_1,\dots,k_n} = W^T h^{(n_h)}_{k_1,\dots,k_n} , \tag{4.15}$$

where W denotes the weights of the linear regression. Therefore, HIR can be learned by minimizing the squared error, and the objective function can be written as:

$$\begin{aligned} J &= \sum (y_{k_1,\dots,k_n} - \hat{y}_{k_1,\dots,k_n})^2 + \frac{\lambda}{2} \|\Theta\|^2 \\ &= \sum (y_{k_1,\dots,k_n} - W^T h^{(n_h)}_{k_1,\dots,k_n})^2 + \frac{\lambda}{2} \|\Theta\|^2 \end{aligned} \tag{4.16}$$

where $\Theta = \{E, T, H, W\}$ denotes all the parameters needed to be learned in HIR, and λ is a parameter to control the power of regularization.

The derivations of J with respect to the parameters can be calculated as:

$$\frac{\partial J}{\partial W} = -2 \sum (y_{k_1,\dots,k_n} - W^T h^{(n_h)}_{k_1,\dots,k_n}) h^{(n_h)}_{k_1,\dots,k_n} + \lambda W ,$$

$$\frac{\partial J}{\partial h^{(n_h)}_{k_1,\dots,k_n}} = -2 \sum (y_{k_1,\dots,k_n} - W^T h^{(n_h)}_{k_1,\dots,k_n}) W + \lambda h^{(n_h)}_{k_1,\dots,k_n} .$$

4.5.2   Ranking Task

Different from the scenario with explicit feedbacks, scenarios with implicit feedback are more common in real-world applications, which form ranking tasks. Bayesian

Personalized Ranking (BPR) is a state-of-the-art pair-wise ranking method for the implicit feedback data. The basic assumption of BPR is that the probability of a positive sample is bigger than that of a negative sample. So, using HIR in the BPR framework, we need to maximize the following function:

$$p(k_n \succ_{k_1,k_2,\dots} k'_n) = g(\hat{y}_{k_1,k_2,\dots,k_n} - \hat{y}_{k_1,k_2,\dots,k'_n}) , \qquad (4.17)$$

where k_n denotes a positive sample, k'_n denotes a negative sample, and $g(x)$ is a nonlinear function which we select as $g(x) = 1/(1 + exp(-x))$. Note that the prediction here is also based on linear regression as shown in Eq. (4.15). Incorporating negative log likelihood, we can solve the following objective function equivalently:

$$\begin{aligned} J &= -\sum \ln(p(k_n \succ_{k_1,k_2,\dots} k'_n)) + \frac{\lambda}{2} \|\Theta\|^2 \\ &= -\sum \ln\left(\frac{1}{1 + \exp(-y(k_n \succ_{k_1,k_2,\dots} k'_n))}\right) + \frac{\lambda}{2} \|\Theta\|^2 \\ &= \sum \ln(1 + \exp(-y(k_n \succ_{k_1,k_2,\dots} k'_n))) + \frac{\lambda}{2} \|\Theta\|^2 \\ &= \sum \ln(1 + \exp(-W^T h(k_n \succ_{k_1,k_2,\dots} k'_n))) + \frac{\lambda}{2} \|\Theta\|^2 \end{aligned} \qquad (4.18)$$

where

$$y(k_n \succ_{k_1,k_2,\dots} k'_n) = \hat{y}_{k_1,k_2,\dots,k_n} - \hat{y}_{k_1,k_2,\dots,k'_n} ,$$

$$h(k_n \succ_{k_1,k_2,\dots} k'_n) = h^{(n_h)}_{k_1,k_2,\dots,k_n} - h^{(n_h)}_{k_1,k_2,\dots,k'_n} .$$

The derivations of J with respect to the parameters can be calculated as:

$$\frac{\partial J}{\partial W} = -\sum q(k_n \succ_{k_1,k_2,\dots} k'_n)h(k_n \succ_{k_1,k_2,\dots} k'_n) + \lambda W ,$$

$$\frac{\partial J}{\partial h^{(n_h)}_{k_1,k_2,\dots,k_n}} = -\sum q(k_n \succ_{k_1,k_2,\dots} k'_n)W + \lambda h^{(n_h)}_{k_1,k_2,\dots,k_n} ,$$

$$\frac{\partial J}{\partial h^{(n_h)}_{k_1,k_2,\dots,k'_n}} = \sum q(k_n \succ_{k_1,k_2,\dots} k'_n)W + \lambda h^{(n_h)}_{k_1,k_2,\dots,k'_n} ,$$

where

$$q(k_n \succ_{k_1,k_2,\dots} k'_n) = \frac{\exp(-W^T h(k_n \succ_{k_1,k_2,\dots} k'_n))}{1 + \exp(-W^T h(k_n \succ_{k_1,k_2,\dots} k'_n))} .$$

4.5.3 Classification Task

In some scenarios, we need to predict the interaction probability of the multiple entities and contexts, e.g., click-through rate prediction. It can be treated as a classification task. We incorporate logistic regression, and the prediction becomes:

$$\hat{y}_{k_1,\dots,k_n} = \frac{1}{1 + exp(-W^T h^{(n_h)}_{k_1,\dots,k_n})} , \tag{4.19}$$

where W denotes the weights of the logistic regression. As in regular logistic regression, we use negative log likelihood for the model learning. The objective function can be written as:

$$\begin{aligned}
J &= -\sum \left(1 - y_{k_1,k_2,\dots,k_n}\right) \ln \left(1 - \hat{y}_{k_1,k_2,\dots,k_n}\right) \\
&\quad - \sum y_{k_1,k_2,\dots,k_n} \ln \left(\hat{y}_{k_1,k_2,\dots,k_n}\right) + \frac{\lambda}{2} \|\Theta\|^2 \\
&= \sum \ln(1 + \exp(-W^T h^{(n_h)}_{k_1,\dots,k_n})) \\
&\quad + \sum (1 - y_{k_1,k_2,\dots,k_n}) W^T h^{(n_h)}_{k_1,\dots,k_n} + \frac{\lambda}{2} \|\Theta\|^2
\end{aligned} \tag{4.20}$$

The derivations of J with respect to the parameters can be calculated as:

$$\begin{aligned}
\frac{\partial J}{\partial W} &= -\sum \frac{\exp(-W^T h^{(n_h)}_{k_1,\dots,k_n})}{1 + \exp(-W^T h^{(n_h)}_{k_1,\dots,k_n})} h^{(n_h)}_{k_1,\dots,k_n} \\
&\quad + \sum (1 - \hat{y}_{k_1,k_2,\dots,k_n}) + \lambda W
\end{aligned} ,$$

$$\begin{aligned}
\frac{\partial J}{\partial h^{(n_h)}_{k_1,\dots,k_n}} &= -\sum \frac{\exp(-W^T h^{(n_h)}_{k_1,\dots,k_n})}{1 + \exp(-W^T h^{(n_h)}_{k_1,\dots,k_n})} W \\
&\quad + \sum (1 - \hat{y}_{k_1,k_2,\dots,k_n}) + \lambda h^{(n_h)}_{k_1,\dots,k_n}
\end{aligned} .$$

4.6   Iterative Parameter Learning

In three tasks mentioned above, the derivation $\dfrac{\partial J}{\partial h^{(n_h)}_{k_1,\dots,k_n}}$ has been calculated. Based on this derivation, we can calculate the derivations of hidden layers and joint representation iteratively. Suppose we have $\dfrac{\partial J}{\partial h^{(m_h)}_{k_1,\dots,k_m}}$ of the m_hth ($0 < m_h \le n_h$) hidden layer, we can calculate the derivations of parameters on this layer as:

$$\frac{\partial J}{\partial H_{m_h}} = \frac{\partial J}{\partial h^{(m_h)}_{k_1,k_2,\dots,k_n}} h^{(m_h-1)}_{k_1,k_2,\dots,k_n} \;,$$

$$\frac{\partial J}{\partial h^{(m_h-1)}_{k_1,k_2,\dots,k_n}} = H^T_{m_h} \frac{\partial J}{\partial h^{(m_h)}_{k_1,k_2,\dots,k_n}} \;.$$

On the first hidden layer, we can obtain the derivation of the joint representation:

$$\frac{\partial J}{\partial r^{(n)}_{k_1,\dots,k_n}} = \frac{\partial J}{\partial h^{(0)}_{k_1,\dots,k_n}} \;.$$

Now, with the derivation of the joint representation $\dfrac{\partial J}{\partial r^{(n)}_{k_1,\dots,k_n}}$, we can calculate all the derivations iteratively. Suppose we have $\dfrac{\partial J}{\partial r^{(m)}_{k_1,\dots,k_m}}$ of the mth ($1 < m \le n$) layer, we can calculate the derivations of parameters on this layer as:

$$\frac{\partial J}{\partial r^{(m-1)}_{k_1,\dots,k_{m-1}}} = [\left(e^{(m)}_{k_m}\right)^T T^{(m-1)}]^T \frac{\partial J}{\partial r^{(m)}_{k_1,\dots,k_n}} \;,$$

$$\frac{\partial J}{\partial e^{(m)}_{k_m}} = [(r^{(m-1)}_{k_1,\dots,k_{m-1}})^T (T^{(m-1)})^T]^T \frac{\partial J}{\partial r^{(m)}_{k_1,\dots,k_n}} \;,$$

$$\frac{\partial J}{\partial T^{(m-1)}_i} = e^{(m)}_{k_m} \left(r^{(m-1)}_{k_1,\dots,k_{m-1}}\right)^T \left(\frac{\partial J}{\partial r^{(m)}_{k_1,\dots,k_n}}\right)_i \;.$$

On the first layer, we can obtain the derivation of the first entity:

$$\frac{\partial J}{\partial e^{(1)}_{k_1}} = \frac{\partial J}{\partial r^{(1)}_{k_1}} \;.$$

After calculating all the derivations, a solution of HIR can be obtained by using stochastic gradient descent (SGD). Note that the learning method can be changed according to different applications.

4.7 Conclusion

This chapter introduces a novel method, i.e., HIR, for modeling multi-entity interaction. HIR generates the interaction representation of two entities via tensor multiplication, and this process is repeated iteratively for the interaction of several entities and contexts. This procedure forms a hierarchical structure and can generate the final joint representation. Then the collaborative prediction can be calculated based on this representation using various learning methods.

References

1. Liu, Q., Wu, S., Wang, L.: Collaborative prediction for multi-entity interaction with hierarchical representation. In: CIKM, pp. 613–622 (2015)
2. Mnih, A., Salakhutdinov, R.: Probabilistic matrix factorization. In: Proceedings on Neural Information Processing systems (2007)
3. Rendle, S.: Factorization machines with libfm. ACM Trans. Intell. Syst. Technol. (TIST) **3**(3), 57 (2012)
4. Bengio, Y., Courville, A., Vincent, P.: Representation learning: a review and new perspectives. IEEE Trans. Pattern Anal. Mach. Intell. **35**(8), 1798–1828 (2013)

Chapter 5
Context-Aware Recurrent Structure

Abstract To investigate and address the problem of context-aware sequential pre-diction, this chapter introduces a sequential prediction model, named context-aware recurrent neural networks (CA-RNNs). Instead of using the constant input matrix and transition matrix in conventional RNN models, CA-RNN uses a context-aware recurrent structure and employs context-specific input matrices and context-specific transition matrices. The context-specific input matrix captures the situations that user behaviors happen, such as time, location, weather, while the context-specific transition matrix captures the time intervals between adjacent behaviors in historical sequences.

5.1 Introduction

For sequential prediction, Markov Chain (MC)-based methods [1, 4, 5] have been widely used, which aim to predict the users' next behavior based on the past behaviors in sequential data. A transition matrix is estimated, which can give the probability of an action based on the previous ones. For collaborative prediction, Factorizing Per-sonalized Markov Chain (FPMC) provides more accurate prediction by factorizing a personalized transition tensor. A major problem of MC-based methods is that all the components are independently combined, in other words, these methods make strong independence assumption among multiple factors. To alleviate this limitation, as a typical representation learning model, recurrent neural networks (RNNs) have been employed to model temporal dependency. It achieves state-of-the-art performances in different applications, e.g., sentence modeling tasks [3], location prediction [2], and next basket recommendation [6]. RNN consists of an input layer, an output unit, and multiple hidden layers. The hidden can change dynamically along with the sequential history. Each layer of RNN contains an input element and recurrent transition from the previous status, which are captured by an input matrix and a transition matrix, respectively.

Parts of this chapter is reprinted from [2], with permission from AAAI.

S. Wu et al., *Context-Aware Collaborative Prediction*, SpringerBriefs
in Computer Science, https://doi.org/10.1007/978-981-10-5373-3_5

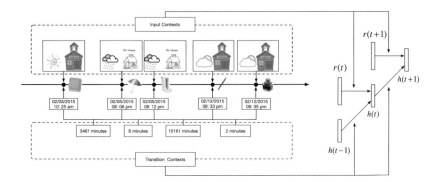

Fig. 5.1 The purchasing sequence of a user is treated as an example of context-aware sequential prediction. The left part shows input contexts and transition contexts in a behavioral sequence. Input contexts mean external situations that users conduct behaviors and transition contexts denote time intervals between adjacent behaviors. The right part illustrates how input contexts and transition contexts contribute to predicting a user's next behaviors in recurrent neural networks

As shown in Fig. 5.1, complex real-world applications usually have not only sequential information but also contains a large amount of contextual information. However, existing sequential prediction methods and context-aware collaborative predictions both have difficulty in dealing with this practical situation. Though RNN and other sequential methods have achieved satisfactory performances in sequential prediction, they still have the difficulty in modeling the rich contextual information. On the other hand, context-aware collaborative prediction has been extensively studied and several methods have been proposed to achieve state-of-the-art performances. But these context-aware prediction methods cannot take sequential information into consideration.

To construct a model to capture the sequential information and contextual information simultaneously, we first investigate the properties of sequential behavioral histories. Here, we conclude two types of contexts, i.e., input contexts and transition contexts, which are demonstrated in Fig. 5.1. Input contexts denote the contexts where input elements in behavioral sequences happen, that is to say, input contexts are external situations that users conduct behaviors, e.g., shopping, visiting, or reading. Such contexts usually include location (home or working place), time (weekdays or weekends, morning or evening), weather (sunny or rainy). Transition contexts are the contexts of the transition from previous behaviors to future behaviors in historical sequences. Specifically, transition contexts denote time intervals between adjacent behaviors in sequences. Generally speaking, long time intervals and short time intervals mean great different for the transition from the past.

We introduce context-aware recurrent neural networks (CA-RNNs) to model sequential information and contextual information in one framework. Instead of using a constant input matrix to capture input elements in each layer of RNN, we use context-specific input matrices for each specific input contexts. Similarly, we use context-specific transition matrices for modeling transition effects from previous behaviors in historical sequences under specific transition contexts, i.e., time intervals between adjacent behaviors. Then, we implement our CA-RNN model in a Bayesian personalized ranking framework to make personalized ranking of recommended items, and backpropagation through time is applied for learning parameters of CA-RNN.

In this chapter, we first introduce the definition and notations of context-aware sequential prediction. Then, we show how to model input contexts and transition contexts respectively. Finally, we formulate the prediction function and introduce the parameter learning process of the CA-RNN model.

5.2 Notations

Here, we first introduce the problem definition. We have a set of users denoted as $U = \{u_1, u_2, ...\}$, a set of items denoted as $V = \{v_1, v_2, ...\}$. There are multiple types of input contexts denoted as $C_I = \{C_{i1}, C_{i2}, ...\}$, where C_{ik} denotes one specific type of input context such as weather, location. Meanwhile, each input context contains several values, e.g., the weather context may consist of sunny, rainy, cloudy and so on. C_T denotes transition contexts, i.e., time intervals between two adjacent behaviors in behavioral history.

For each user u, the behavioral history is given as $V^u = \{v_1^u, v_2^u, ...\}$, where v_k^u denotes the kth selected item of user u. And the behavioral history of all users is denoted as $V^U = \{V_1^u, V_2^u, ...\}$. Each behavior in the history is associated with corresponding timestamps $T^u = \{t_1^u, t_2^u, ...\}$, where t_k^u denotes the kth timestamp in the behavioral sequence of user u. For each user u, at specific timestamp t_k^u, a combination of input contexts is denoted as $c_{I,k}^u = \{c_{i1,k}^u, c_{i2,k}^u, ...\}$, which consists of several types of input contexts. And the corresponding transition context is denoted as $c_{T,k}^u$, which is decided by the time interval between the timestamp t_k^u of the current behavior and the timestamp t_{k-1}^u of the previous behavior.

Given the behavioral history $V^u = \{v_1^u, v_2^u, ..., v_k^u\}$ containing input and transition contexts of user u till timestamp t_k^u, we would like to predict the user's selected item v_{k+1}^u under input contexts $c_{I,k+1}^u$ and transition contexts $c_{T,k+1}^u$.

Though RNN has achieved successful performances in sequential modeling, it has difficulty in modeling a variety of contextual information. With the increasing of contextual information in practical applications, context-aware sequential prediction becomes an emerging task. As a consequence, it is appropriate to incorporate significant contextual information of sequence into an adapted RNN architecture.

5.3 Context Modeling

5.3.1 Recurrent Neural Networks

The architecture of a recurrent neural network consists of an input layer i, a hidden layer h, an output unit, as well as inner weight matrices. RNN model the current output as the function of the previous output and a hidden layer. At each time step, we can predict the output unit given the hidden layer, and then feed the new output back into the next hidden state. The formulation of the hidden layer in RNN is:

$$\mathbf{h}_k^u = f\left(\mathbf{r}_{v_k^u}\mathbf{M} + \mathbf{h}_{k-1}^u\mathbf{W}\right), \tag{5.1}$$

where $\mathbf{h}_k^u$ is the d-dimensional representation of user u at timestamp t_k^u in a behavior sequence, $\mathbf{r}_{v_k^u}$ denotes the d-dimensional latent vector of the corresponding item the user selects, $\mathbf{W}$ is the $d \times d$ dimensional recurrent connection of the previous status propagating sequential signals, and $\mathbf{M}$ denotes the $d \times d$ dimensional transition matrix for input elements to capture the current behavior of the user. The activation function $f(x)$ is usually chosen as a *sigmoid* function $f(x) = \exp\left(1/1 + e^{-x}\right)$.

5.3.2 Modeling Input Contexts

Input contexts denote the external situations that users conduct behaviors. They have significant effects on user behaviors and are essential for predicting the future. Thus, we also incorporate input contexts in conventional RNN model. The constant input matrix in conventional RNN is replaced with context-specific input matrices according to different input contexts. Then, Eq. 5.1 can be rewritten as:

$$\mathbf{h}_k^u = f\left(\mathbf{r}_{v_k^u}\mathbf{M}_{c_{I,k}^u} + \mathbf{h}_{k-1}^u\mathbf{W}\right), \tag{5.2}$$

where $c_{I,k}^u$ is the combination of input contexts at timestamp t_k^u of user u and $\mathbf{M}_{c_{I,k}^u}$ denotes the corresponding $d \times d$ dimensional context-aware input matrix. This captures how external situations affects future behaviors of users.

   Moreover, there are usually multiple types of input contexts in one situation, e.g., weather, location, and mood. It is necessary to aggregate the matrices of each type of input contexts $c_{i,k}^u$ to generate the context-specific input matrix of the input context combination $c_{I,k}^u$. We study three typical aggregation methods as follows:

- Linear aggregation method: Linear computation is the most common method to calculate the aggregation of all context-specific matrices. It treats the effects of each matrix separately, and the $\mathbf{M}_{c_{I,k}^u}$ can be calculated as:

$$\mathbf{M}_{c_{I,k}^u} = \sum_{c_{i,k}^u \in C^I} \mathbf{M}_{c_{i,k}^u} . \qquad (5.3)$$

- Multiplied aggregation method: The work of [2] incorporates temporal and spatial contexts into conventional RNN by multiplying time-specific operating matrix and distance-specific operating matrix. Analogously, we can treat input matrices of different types of input contexts in the same way:

$$\mathbf{M}_{c_{I,k}^u} = \prod_{c_{i,k}^u \in C^I} \mathbf{M}_{c_{i,k}^u} . \qquad (5.4)$$

- Combination-specific aggregation method: Other than using linear method or multiplied method, we can also simply associate each input context combination with a specific matrix. It means each combination of input contexts has its own input matrix, and we do not have matrices for each type of input contexts. In most cases, this will bring more parameters to be estimated.

5.3.3 Modeling Transition Contexts

Different length of time interval between two behaviors has different impacts on prediction of the future. Longer time intervals usually have limited effect in predicting next behavior comparing with shorter ones. Therefore, the length of time interval is essential for predicting future behaviors. Thus, we conclude time intervals as transition contexts, and incorporate transition contexts in conventional RNN model. The constant transition matrix in conventional RNN is replaced with context-specific transition matrices according to different transition contexts. And Eq. 5.1 should be rewritten as:

$$\mathbf{h}_k^u = f\left(\mathbf{r}_{v_k^u}\mathbf{M} + \mathbf{h}_{k-1}^u \mathbf{W}_{c_{T,k}^u}\right) , \qquad (5.5)$$

where $\mathbf{W}_{c_{T,k}^u}$ is a $d \times d$ dimensional context-specific transition matrix for the transition context $c_{T,k}^u$. It captures how previous behaviors affects the future. The context-specific transition matrix can be generated according to the time interval between the current timestamp and the previous timestamp:

$$\mathbf{W}_{c_{T,k}^u} = \mathbf{W}_{t_k^u - t_{k-1}^u}, \qquad (5.6)$$

where t_k^u and t_{k-1}^u denotes the current and the previous timestamp, respectively, and $t_k^u - t_{k-1}^u$ is the corresponding time interval.

It is impossible to learn a context-specific transition matrix for every possible continuous time interval. Then, we partition the range of all the possible time intervals into discrete time bins. To better learn all the transition matrices, each time bin should contain approximately the same amount of observations. And the transition matrix

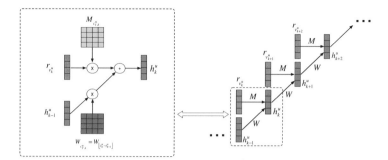

Fig. 5.2 Overview of the CA-RNN model. The right part shows the forward propagation process of CA-RNN. The left part illustrates the computational procedure of a hidden layer in CA-RNN

of each time interval value is computed based on the floor of the corresponding time interval value. So, in practice, context-specific transition matrices can be generated as:

$$\mathbf{W}_{c_{T,k}^u} = \mathbf{W}_{\lfloor t_k^u - t_{k-1}^u \rfloor}. \tag{5.7}$$

Then, as shown in Fig. 5.2 and the formulation of each hidden layer in the CA-RNN model becomes:

$$\mathbf{h}_k^u = f\left(\mathbf{r}_{v_k^u}\mathbf{M} + \mathbf{h}_{k-1}^u \mathbf{W}_{\lfloor t_k^u - t_{k-1}^u \rfloor}\right), \tag{5.8}$$

where $\lfloor t_k^u - t_{k-1}^u \rfloor$ denotes the floor of the corresponding time interval $t_k^u - t_{k-1}^u$.

5.4 Context-Aware Recurrent Neural Networks

5.4.1 Hidden Layer

We have detailed our method to model input contexts and transition contexts. It is reasonable to combine the two kinds of contexts in one framework. Then, we can generate the formulation of each hidden layer of CA-RNN as:

$$\mathbf{h}_k^u = f\left(\mathbf{r}_{v_k^u}\mathbf{M}_{c_{I,k}^u} + \mathbf{h}_{k-1}^u \mathbf{W}_{c_{T,k}^u}\right). \tag{5.9}$$

This formulation can take both input and transition contexts into consideration and make better prediction of future behaviors of users. Moreover, according to the approximation in Eq. 5.7, in practice, the formulation can be rewritten as:

$$\mathbf{h}_k^u = f\left(\mathbf{r}_{v_k^u}\mathbf{M}_{c_{I,k}^u} + \mathbf{h}_{k-1}^u \mathbf{W}_{\lfloor t_k^u - t_{k-1}^u \rfloor}\right). \tag{5.10}$$

5.4.2 Context-Aware Sequential Prediction

When making prediction of user behaviors at timestamp t_{k+1}^u, despite contexts in the behavioral history modeled above, present contexts $c_{I,k+1}^u$ and $c_{T,k+1}^u$ also have significant effects on making prediction. Thus, our method should incorporate present contexts and make context-aware prediction. Therefore, the prediction of whether user u will select item v at timestamp t_{k+1}^u can be computed as:

$$y_{u,k+1,v} = \mathbf{h}_k^u \mathbf{P}_{k+1}^u (\mathbf{r}_v)^\mathsf{T} , \qquad (5.11)$$

where $\mathbf{P}_{k+1}^u$ is the $d \times d$ dimensional matrix representation of input contexts and transition contexts of user u at timestamp t_{k+1}^u. This matrix indicates the effect of present contexts on the next behavior of user u. It can be generated as:

$$\mathbf{P}_{k+1}^u = \mathbf{M}'_{c_{I,k+1}^u} \mathbf{W}'_{c_{T,k+1}^u} , \qquad (5.12)$$

where $\mathbf{M}'_{c_{I,k+1}^u}$ and $\mathbf{W}'_{c_{T,k+1}^u}$ denote $d \times d$ dimensional matrix representations of present input context combination $c_{I,k+1}^u$ and transition context $c_{T,k+1}^u$, respectively, at timestamp t_{k+1}^u. And Eq. 5.13 can be rewritten as:

$$y_{u,k+1,v} = \mathbf{h}_k^u \mathbf{M}'_{c_{I,k+1}^u} \mathbf{W}'_{c_{T,k+1}^u} (\mathbf{r}_v)^\mathsf{T} , \qquad (5.13)$$

where $\mathbf{M}'_{c_{I,k+1}^u}$ can also be generated according to the aggregation methods mentioned above and $\mathbf{W}'_{c_{T,k+1}^u}$ is computed based on the approximation in Eq. 5.7.

5.5 Learning Algorithm

In this subsection, we introduce the learning process of CA-RNN with Bayesian Personalized Ranking (BPR) and Back Propagation Through Time (BPTT). BPR and BPTT has been successfully used for learning of RNN models [2, 6].

BPR is a state-of-the-art pair-wise ranking framework for the implicit feedback data. The basic assumption of BPR is that a user prefers a selected item than a negative one. Thus, the training objective of CA-RNN under the BPR framework is to maximize the following probability:

$$p(u, k+1, v \succ v') = g(y_{u,k+1,v} - y_{u,k+1,v'}), \qquad (5.14)$$

where v' denotes a negative item sample and $g(x)$ is a nonlinear function which is selected as $g(x) = 1/(1 + e^{-x})$. Incorporating the negative log likelihood, we can solve the following objective function equivalently:

$$J = \sum \ln(1 + e^{-(y_{u,k+1,v} - y_{u,k+1,v'})}) + \frac{\lambda}{2}(\|\mathbf{R}\|^2 + \|\mathbf{M}\|^2 + \|\mathbf{W}\|^2) , \qquad (5.15)$$

where $\mathbf{R}, \mathbf{M}, \mathbf{W}$ denote all the parameters to be estimated and λ is a parameter to control the power of regularization. And the derivation of J with respect to the parameters can be calculated as:

$$\frac{\partial J}{\partial \mathbf{r}_v} = -\sum \frac{\mathbf{h}_k^u \mathbf{P}_{k+1}^u e^{-(y_{u,k+1,v}-y_{u,k+1,v'})}}{1+e^{-(y_{u,k+1,v}-y_{u,k+1,v'})}} + \lambda \mathbf{r}_v ,$$

$$\frac{\partial J}{\partial \mathbf{r}_{v'}} = \sum \frac{\mathbf{h}_k^u \mathbf{P}_{k+1}^u e^{-(y_{u,k+1,v}-y_{u,k+1,v'})}}{1+e^{-(y_{u,k+1,v}-y_{u,k+1,v'})}} + \lambda \mathbf{r}_{v'} ,$$

$$\frac{\partial J}{\partial \mathbf{h}_k^u} = -\sum \frac{(\mathbf{r}_v - \mathbf{r}_v')(\mathbf{P}_{k+1}^u)^{\mathrm{T}} e^{-(y_{u,k+1,v}-y_{u,k+1,v'})}}{1+e^{-(y_{u,k+1,v}-y_{u,k+1,v'})}} ,$$

$$\frac{\partial J}{\partial \mathbf{P}_{k+1}^u} = -\sum \frac{(\mathbf{h}_k^u)^{\mathrm{T}}(\mathbf{r}_v - \mathbf{r}_v') e^{-(y_{u,k+1,v}-y_{u,k+1,v'})}}{e^{-(y_{u,k+1,v}-y_{u,k+1,v'})}} .$$

And given $\partial J / \partial \mathbf{P}_{k+1}^u$, we can achieve:

$$\frac{\partial J}{\partial \mathbf{M}_{c_{I,k+1}'}^u} = \frac{\partial J}{\partial \mathbf{P}_{k+1}^u} \left(\mathbf{W}_{c_{T,k+1}'}^u \right)^{\mathrm{T}} ,$$

$$\frac{\partial J}{\partial \mathbf{W}_{c_{T,k+1}'}^u} = \left(\mathbf{M}_{c_{I,k+1}'}^u \right)^{\mathrm{T}} \frac{\partial J}{\partial \mathbf{P}_{k+1}^u} .$$

Moreover, parameters in CA-RNN can be further learnt by using the BPTT algorithm. According to Eq. 5.9, given the derivation $\partial J / \partial \mathbf{h}_k^u$, the corresponding gradients of all parameters in the hidden layer can be calculated as:

$$\frac{\partial J}{\partial \mathbf{h}_{k-1}^u} = \left(f'(\cdot) \otimes \frac{\partial J}{\partial \mathbf{h}_k^u} \right) \left(\mathbf{W}_{c_{T,k}^u} \right)^{\mathrm{T}} ,$$

$$\frac{\partial J}{\partial \mathbf{W}_{c_{T,k}^u}} = \left(\mathbf{h}_{k-1}^u \right)^{\mathrm{T}} \left(f'(\cdot) \otimes \frac{\partial J}{\partial \mathbf{h}_k^u} \right) ,$$

$$\frac{\partial J}{\partial \mathbf{r}_{v_k^u}} = \left(f'(\cdot) \otimes \frac{\partial J}{\partial \mathbf{h}_k^u} \right) \left(\mathbf{M}_{c_{I,k}^u} \right)^{\mathrm{T}} ,$$

$$\frac{\partial J}{\partial \mathbf{M}_{c_{I,k}^u}} = \left(\mathbf{r}_{v_k^u} \right)^{\mathrm{T}} \left(f'(\cdot) \otimes \frac{\partial J}{\partial \mathbf{h}_k^u} \right) .$$

This process can be repeated iteratively in the whole behavioral sequence. After calculating all the gradients, we can employ stochastic gradient descent (SGD) to estimate the model parameters.

5.6   Conclusion

This chapter shows the model of context-aware recurrent neural networks, CA-RNN, which has a context-aware recurrent structure. This model replaces the constant input matrix of conventional RNN with context-specific input matrices, in order to model complex real-world contexts, e.g., time, location, and weather. Meanwhile, to model transition contexts, i.e., time intervals between adjacent behaviors in historical sequences, context-specific transition matrices are incorporated, instead of the constant one in the conventional RNN. Then, to make personalized ranking of recommended items, CA-RNN is implemented under the BPR framework, and BPTT is used to learn parameters in CA-RNN.

References

1. Chen, J., Wang, C., Wang, J.: A personalized interest-forgetting Markov model for recommendations. In: AAAI, pp. 16–22 (2015)
2. Liu, Q., Wu, S., Wang, L., Tan, T.: Predicting the next location: a recurrent model with spatial and temporal contexts. In: AAAI, pp. 194–200 (2016)
3. Mikolov, T., Karafiát, M., Burget, L., Cernockỳ, J., Khudanpur, S.: Recurrent neural network based language model. In: INTERSPEECH, pp. 1045–1048 (2010)
4. Natarajan, N., Shin, D., Dhillon, I.S.: Which app will you use next?: Collaborative filtering with interactional context. In: RecSys, pp. 201–208 (2013)
5. Yang, Q., Fan, J., Wang, J., Zhou, L.: Personalizing web page recommendation via collaborative filtering and topic-aware Markov model. In: ICDM, pp. 1145–1150 (2010)
6. Yu, F., Liu, Q., Wu, S., Wang, L., Tan, T.: A dynamic recurrent model for next basket recommendation. In: SIGIR, pp. 729–732 (2016)

Chapter 6
Performance of Different Collaborative Prediction Tasks

Abstract This chapter contains the experiments of four tasks, i.e., general recommendation, context-aware recommendation, latent collaborative retrieval, and click-through rate prediction. At first, this chapter describes the representative methods of collaborative prediction, context-aware collaborative prediction, and context-aware sequential recommendation. Then, it shows the experimental settings including the datasets and evaluation metrics. The experimental results on real datasets show that COT, HIR, and CA-RNN, respectively, outperform the state-of-the-art methods of context-aware collaborative prediction and context-aware sequential prediction. In addition, in context-aware recommendation and latent collaborative retrieval, we analyze the impact of the dimensionality of latent representations and examine the interacting order of entities. The convergence performance, the scalability, and impact of parameters are also analyzed. Last but not the least, we visualize the representations in latent collaborative retrieval and find some interesting observations on context representations and context weights.

6.1 Collaborative Prediction Methods

To investigate the effectiveness of COT, HIR, and CA-RNN models from various angles, there are several compared methods which should be introduced, which are collaborative prediction, context-aware prediction, context-aware sequential prediction. Many of them are state-of-the-art and broadly used methods for different tasks of collaborative prediction.

6.1.1 Collaborative Prediction

Baseline methods of collaborative prediction include logistic regression, matrix factorization methods, and some extensions of factorization methods.

Parts of this chapter is reprinted from [1–3], with permission from IEEE, ACM, and AAAI.

© The Author(s) 2017   53
S. Wu et al., *Context-Aware Collaborative Prediction*, SpringerBriefs
in Computer Science, https://doi.org/10.1007/978-981-10-5373-3_6

- **POP** is a naive baseline method that predicts the most popular items to users.
- **LR** is a widely used classifier and used as a baseline method for click-through rate prediction [4, 5]. It also can be implemented based on one-hot representation for context-aware recommendation and latent collaborative retrieval.
- **MF** [6] is used as a baseline method for the general recommendation and should ignore the *context* factor in context-aware recommendation and ignore the *user* factor in latent collaborative retrieval.
- **BPR** [7] is the state-of-the-art method for implicit feedback in the general recommendation.
- **SVD++** [8] is an advanced matrix factorization model, but is not designed for the context-aware recommendation. We implement it as a baseline of the general recommendation in our experiments.
- **FM** [9] is applicable for different kinds of contextual information by specifying the input data. We use LibFM[1] to implement this general method.

6.1.2 Context-Aware Collaborative Prediction

Context-aware collaborative prediction can make prediction by taking the contextual information into consideration. Major of these models are constructed based on the factorization method.

- **TF** [10] is a state-of-the-art method for context-aware recommendation and tag recommendation. We use it as a baseline method for both context-aware recommendation and latent collaborative retrieval.
- **Multiverse Recommendation** [11] is a state-of-the-art model which employs Tucker decomposition on the user-item-context rating tensor. This model outperforms conventional context-aware recommendation models, such as the prefiltering and multidimensional approach.
- **HeteroMF** [12] is a multi-domain relation prediction method which uses transfer matrices and can be used to model the contextual information. Each specific context combination has a transfer matrix.
- **CARS2** [13] provides each user and item with a latent vector and a context-aware representation. The context-aware representation captures latent properties of the user and item manipulated by the contextual information. It has achieved the state-of-the-art performance in context-aware personalized ranking.
- **LCR** [14] is a state-of-the-art method for latent collaborative retrieval.

---

[1] http://www.libfm.org/.

6.1.3 Sequential Collaborative Prediction

Sequential information is a significant factor for collaborative prediction; various sequential collaborative prediction methods have been proposed for practical systems. They can be categorized into Markov chain methods and RNN-based methods.

- **FPMC** [15] extends conventional Markov chain methods and factorizes personalized probability transition matrices. It is a widely used method for sequential prediction and recommendation.
- **HRM** [16] learns the representation of behaviors from the previous transactions and predicts next behaviors of users. It has become a state-of-the-art method for next basket recommendation.
- **RNN** [17] is a state-of-the-art method for the sequential prediction, which has been successfully applied in various sequential applications.

To provide comparison among different kinds of contexts and different aggregation methods, we implement not only the CA-RNN model, but also some variations of it. To investigate properties of input contexts, we implement CA-RNN-week and CA-RNN-month, which incorporate only the context *week* and *month* respectively. Besides, to find the best method for aggregating the two types of input contexts, we present the performance results of CA-RNN-lin, CA-RNN-mul and CA-RNN-com, which are based on linear aggregation method, multiplied aggregation method and combination-specific aggregation method, respectively. To compare the effects of input contexts and transition contexts, we implement CA-RNN-input and CA-RNN-transition, where the context-aware recurrent structure is with only input contexts or transition contexts.

6.2 Experimental Setting

In this part, first, we introduce the experimental datasets for different tasks of context-aware collaborative prediction and context-aware sequential prediction. Then, we list different kinds of evaluation metrics for regression, ranking, and classification tasks.

6.2.1 Datasets

6.2.1.1 Datasets for Context-Aware Collaborative Prediction

Although the context-aware collaborative prediction is a practical problem, there are only a few publicly available datasets. We investigate the performance of the COT and HIR models on three benchmark datasets: Food, Adom, MovieLens, Delicious and Avazu datasets. The experiments are conducted on these datasets to evaluate the performance of different tasks.

- **Food DataSet** [18] is collected from a restaurant. There are two contexts: $virtuality$ describes if the situation in which the user rates are $virtual$ or $real$, and $hunger$ S captures how hungry the user is. It is a suitable dataset for context-aware recommendation and has also been used in the previous work [11, 13].

- **Adom Dataset** [19] is collected from a movie Web site and has rich contextual information. There are five contexts: $companion$ captures whom the user watches the movie with, $when$ shows whether the user watches the movie at weekend, $release$ indicates whether the user watches the movie on the release weekend, rec captures how the user will recommend the movie, and $where$ indicates whether the user watches the movie in a theater.

- **MovieLens-1M and 10M**[2] is collected from a personalized movie recommender system.[3] There is no explicit contextual information, but the timestamp can be split into two interaction contexts: $hour$ and day. Besides, this dataset contains user and item contexts, i.e., $gender$, age and $occupation$ of the user and $title$ and $genre$ of the item. MovieLens-10M is a widely used dataset for rating prediction in recommender systems, which can be used for general recommendation.

- **Delicious dataset**[4] contains three entities: $user$, URL, and tag, with $0.4M$ observations. This dataset can be used for latent collaborative retrieval when considering URL as query and tag as document, which means a $user$ retrieves $tags$ for an untagged URL. Similar to previous work, we use p-core[5] for filtering the dataset and p is chosen as 10.

- **Avazu dataset**[6] is a dataset for click-through rate prediction, in which we extract five entities: $device$, $Web\ site$, $application$, $advertisement$, and $position$, with $24M$ observations. We conduct two experiments on this dataset with the only difference of whether the entity $position$ is included.

For context-aware recommendation, we use Food, Adom, and MovieLens-1M for experiments. The main difference between Food and Adom lies on the amount of contextual information, which gives us an opportunity to estimate the relation between the model performance and the scale of contexts. The MovieLens-1M dataset is another widely used dataset, where the timestamp can be used as interaction contexts and attributes of users and items can be treated as entity contexts. We employ this dataset to examine the performance of methods in dealing with general contextual information. For all the datasets, we use 70% for training, 20% for testing, and the remaining 10% data as the validation set for tuning parameters, i.e., the dimensions of latent representations. And the regulation parameter is set as $\lambda = 0.01$.

---

[2]http://grouplens.org/datasets/movielens/.

[3]http://movielens.org/.

[4]http://grouplens.org/datasets/hetrec-2011/.

[5]The p-core of the dataset is the largest subset of the dataset with the property that every user, every item and every tag has to occur in at least p posts.

[6]https://www.kaggle.com/c/avazu-ctr-prediction/data.

6.2.1.2 Datasets for Context-Aware Sequential Prediction

To assess the performance of context-aware sequential prediction, our experiments are conducted on two real sequential datasets with rich contextual information:

- **Tmall**[7] is a public dataset collected from Tmall,[8] one of the biggest e-commerce Web sites in China. It covers products from food, clothes to electrical appliance. The dataset records online purchasing and clicking behaviors of users on various shops. It contains 182,881 records belonging to 885 users and 9,532 shops. The Tmall dataset associates with timestamps of user behaviors, which can be used as contextual information. The temporal information in the dataset is accurate to the day level.
- **Fashion**[9] is a public online e-commerce dataset released by TIANCHI.[10] It records shopping behaviors on fashion products, including clothes, bags, hats. It contains 13,611,038 records belonging to 1,103,702 users and 462,008 items. It also contains the contextual information timestamps, which are collected based on the day level.

To avoid sparsity of users and items of these two datasets, we conduct some data pre-processing. For the Tmall dataset, we remove users who bought less than 10 times and shops that are bought less than 10 times. For the Fashion dataset, we remove users who bought less than 10 items and items that are bought less than 3 times. Because shops in the Tmall dataset are bought more frequently than items in the Fashion dataset, we set the threshold in Tmall larger than that in Fashion. Moreover, for each behavioral sequence of the two datasets, we use first 90% elements in the sequence for training, the remaining 10% data for testing. And the regulation parameter in our experiments is set as $\lambda = 0.01$.

For context-aware sequential prediction, we should first extract input and transition contexts from the datasets. According to the timestamps in the Tmall and Fashion datasets, we extract input contexts and transition contexts to implement the CA-RNN model, as well as other context-aware methods.

First, based on timestamps, we can extract two types of input contexts: different days in a week and different time periods in a month. Users' behaviors are different in different days in a week, especially between weekdays and weekends. So we extract such kind of contexts, which is denoted as context *week*. And salary receiving condition will also affect users' behaviors. For example, people tend to buy clothes and entertainment products in the beginning of the month when receiving salaries and buy necessities of life at the end of the month when lacking money. So we divide a month into three ten-day time periods and denote this kind of contexts as context *month*.

---

[7]https://102.alibaba.com/competition/addDiscovery/index.htm.

[8]https://www.tmall.com/.

[9]https://tianchi.shuju.aliyun.com/competition/index.htm.

[10]https://tianchi.shuju.aliyun.com/.

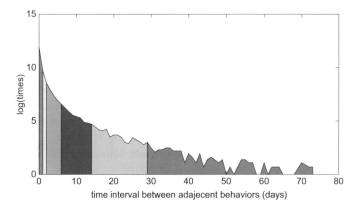

(a) Distribution of time intervals on the Tmall dataset.

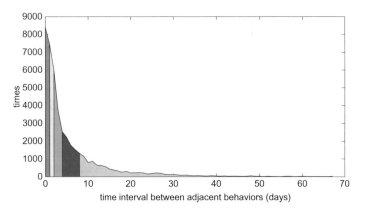

(b) Distribution of time intervals on the Fashion dataset.

Fig. 6.1 Distribution of length of time intervals between adjacent behaviors in sequences on the Tmall and Fashion datasets. To make each time bin contains approximately the same amount of observation, we sign each region with the same color in the figure as a time bin

Second, we extract time intervals between adjacent behaviors in sequences as transition contexts. As we discuss above, we should partition the range of all the possible time intervals into discrete bins. And each bin should contain approximately the same amount of observation. Figure 6.1 illustrates the distribution of length of time intervals between adjacent behaviors in sequences on the Tmall and Fashion datasets. Our partition is done according to the regions with different colors in these figures, and we sign each region with the same color as a time bin. Thresholds for partitioning the Tmall dataset are 0, 1, 2, 7, 15, and 30. And thresholds for partitioning the Fashion dataset are 0, 1, 2, 4, and 8.

6.2.2 Evaluation Metrics

We have different evaluation metrics for different tasks.

- **Root Mean Square Error (RMSE)** and **Mean Average Error (MAE)** are the most popular metrics for general recommendation and context-aware recommendation. For these two metrics, the smaller the value, the better the performance.
- **Recall@k**, **Precision@k** and **F1-score@k** are three important metrics for ranking tasks. We use them for latent collaborative retrieval. The evaluation score for a given *user-URL-tag* triple is computed according to where the tag appears in the ranked list. The larger the value, the better the performance.
- **Mean Average Precision (MAP)** is another metric for the evaluation in latent collaborative retrieval. MAP is a standard metric for evaluating the quality of ranked lists, and its top-bias property is particularly important for ranking tasks. The larger the value, the better the performance.
- **LogLoss** is commonly used for the evaluation in click-through rate prediction. The smaller the value, the better the performance.

6.3 Performance Comparison of Different Tasks

As shown in Table 6.1, we conduct experiments in four application scenarios: general recommendation, context-aware recommendation, latent collaborative retrieval, and click-through rate prediction.

Table 6.1 Experimental summarization

Applications	Tasks	Datasets	Entities	Compared methods	Evaluation metrics
Recommendation	Regression	Movielen-10M	User, movie	MF, FM	RMSE MAE
Context-aware recommendation	Regression	Food dataset	User, item, context	MF, LR, TF, FM, CARS2, COT	RMSE, MAE
Lantent collaborative retreval	Ranking	Delicious dataset	User, url, tag	MF, LR, TF, FM, LCR	recall@5, recall@10, recall@20, MAP
Click-through rate prediction	Classification	Avazu dataset	Device, application, Web site, advertisement, (position)	LR, FM	LogLoss

Table 6.2 Experimental performance of general recommendation evaluated by RMSE and MAE

Dataset	Method	RMSE	MAE
MovieLens-10M	MF	0.8946	0.7159
	FM	0.8912	0.7128
	COT	0.8946	0.7159
	HIR	0.8863	0.7092
	HIR+	0.8814	0.7054
	HIR++	**0.8772**	**0.7025**

6.3.1 General Recommendation

Table 6.2 illustrates the results measured by RMSE and MAE on MovieLens-10M. It shows that HIR achieves the best results in general recommendation compared with MF and FM. Without the contextual information, COT obtains the same performance as the MF method. HIR, HIR+, HIR++ improve the performance of FM by 0.6%, 1.1%, and 1.5%, respectively. The results show that the high-order calculation of tensor multiplication in HIR can better represent the joint characteristics of *user-item* pair and improve the performance of collaborative prediction. This is the merit of the high-order interaction of HIR.

6.3.2 Context-Aware Recommendation

Table 6.3 shows the results of context-aware recommendation measured by RMSE and MAE. The new context-aware methods COT and CARS2 obtain better performance than that of conventional context-aware methods, i.e., TF and FM. HIR outperforms the state-of-the-art context-aware methods. And from the results, we can see that comparing with COT, the performance improvements of HIR, HIR+, and HIR++ are 2.7%, 3.8%, and 4.4%, respectively. The improvement involved by the first and second hidden layer are 1.1% and 0.6%, respectively, which proves that hidden layers can improve the performance but yield smaller improvement than that of HIR itself. Moreover, we can observe that the improvement of hidden layers decreases gradually with its number, which means that the effort of hidden layers has its limitation and we do not need to set too many hidden layers.

6.3.3 Latent Collaborative Retrieval

Table 6.4 illustrates the results of latent collaborative retrieval measured by recall, precision, F1-score, and MAP on the Delicious dataset. Note that all the methods

Table 6.3 Performance of context-aware recommendation evaluated by RMSE and MAE

Dataset	Method	RMSE	MAE
Food dataset	MF	1.1552	0.9484
	LR	1.1263	0.9136
	TF	1.0635	0.8415
	FM	1.0554	0.8453
	CARS2	1.0201	0.8162
	COT	1.0019	0.7921
	HIR	0.9757	0.7816
	HIR+	0.9643	0.7723
	HIR++	**0.9586**	**0.7684**

Table 6.4 Performance of latent collaborative retrieval on the Delicious dataset evaluated by recall, precision, and F1-score

Method	Recall			Precision			F1-score			MAP
	@5	@10	@20	@5	@10	@20	@5	@10	@20	
MF	0.1116	0.1552	0.1952	0.0915	0.0636	0.0401	0.1006	0.0902	0.0665	0.0919
LR	0.1397	0.1963	0.2576	0.1146	0.0804	0.0527	0.1259	0.1141	0.0875	0.1051
TF	0.1613	0.2246	0.2931	0.1323	0.0918	0.0601	0.1454	0.1303	0.0997	0.1144
FM	0.1638	0.2268	0.2965	0.1343	0.0927	0.0607	0.1476	0.1316	0.1008	0.1155
LCR	0.1654	0.2297	0.3013	0.1356	0.0939	0.0617	0.1491	0.1333	0.1024	0.1164
HIR	0.1792	0.2484	0.3274	0.1469	0.1017	0.0671	0.1615	0.1443	0.1114	0.1222
HIR+	0.1856	0.2553	0.3282	0.1522	0.1046	0.0672	0.1672	0.1484	0.1116	0.1255
HIR++	**0.1867**	**0.2565**	**0.3314**	**0.1531**	**0.1049**	**0.0679**	**0.1682**	**0.1489**	**0.1127**	**0.1266**

are implemented under the BPR framework. On recall@5, HIR, HIR+, and HIR++ improve the performance of LCR by 8.5%, 12.1%, and 12.7%, respectively. And on recall@10 and recall@20, the corresponding improvements become 8.3%, 11.4%, 11.8% and 8.6%, 8.9%, 10.1%. Measured by precision and F1-score, the same improvements can be observed. Moreover, on the global evaluation metric MAP, proposed methods HIR, HIR+, HIR++ improve the performance by 5.0%, 7.8%, and 8.8%, respectively. The results obviously show that HIR greatly outperform the compared methods. We can draw the similar conclusion that the performance improvement of hidden layers decreases gradually with the number of hidden layers in most experiments. The table also shows that compared with the results on @5 and @10, the second hidden layer of HIR++ on @20 brings a greater improvement, which means that when we generate more predicted results, more hidden layers can be used to improve the performance further.

Table 6.5 Performance of click-through rate prediction evaluated by LogLoss

Dataset	Method	Position	
		Excluded	Included
Avazu dataset	LR	0.4186	0.4145
	FM	0.4053	0.4025
	HIR	0.3994	0.3962
	HIR+	0.3962	0.3934
	HIR++	**0.3953**	**0.3921**

6.3.4   Click-Through Rate Prediction

The results of click-through rate prediction measured by LogLoss are illustrated in Table 6.5. Note that all the methods are implemented with the same objective function under LogLoss. We can observe that both latent factor-based methods, i.e., FM and HIR, outperform LR, which shows that latent factor-based methods can better discover underlying relation of different entities and contexts. Moreover, in both experiments, without or with the *position* context in the advertisement impression, HIR achieves the best performance. Compared with FM, HIR, HIR+, and HIR++ improve the performance by 1.4%, 2.2%, 2.5% and 1.5%, 2.3%, 2.5%, respectively, in these two experiments, which shows the promising performance of HIR for interaction prediction of entities and contexts.

6.3.5   Sequential Recommendation

To evaluate the performance of context-aware sequential prediction, we illustrate the performance comparison of CA-RNN and other compared methods. Table 6.6 illustrates the performance of these methods with dimensionality $d = 10$ on the Tmall and Fashion datasets evaluated by Recall, F1-score, and MAP.

From the results, we can achieve following observations. Comparing with the baseline performances of POP and MF, the context-aware methods FM and CARS2 achieve significant improvement. With contextual operation, CARS2 has a slightly better performance than FM and is a better method for modeling contextual information. Jointly modeling sequential information and collaborative information, the sequential methods FPMC and HRM achieve another great improvement on the two datasets. This indicates that comparing with contextual information, sequential information is more important and has more significant effects on user behaviors. And another sequential method, RNN, achieves the best performance among all the compared methods. Moreover, we can observe that our proposed CA-RNN greatly outperforms RNN on the Tmall and Fashion datasets in terms of all the metrics. Comparing with RNN, CA-RNN relatively improves Recall@1, @5, @10, and MAP

Table 6.6 Performance comparison on the Tmall and Fashion datasets with dimensionality $d = 10$

	Method	Recall@1	Recall@5	Recall@10	F1-score@1	F1-score@5	F1-score@10	MAP
Tmall	POP	0.0169	0.0944	0.1938	0.0169	0.0315	0.0352	0.0763
	BPR	0.0425	0.1436	0.2798	0.0425	0.0479	0.0509	0.1218
	FM	0.0584	0.1641	0.3018	0.0584	0.0547	0.0549	0.1448
	CARS2	0.0616	0.1697	0.3106	0.0616	0.0566	0.0565	0.1496
	FPMC	0.0866	0.2253	0.3405	0.0866	0.0751	0.0619	0.1811
	HRM	0.0956	0.2488	0.3703	0.0956	0.0829	0.0673	0.2001
	RNN	0.1282	0.3412	0.4392	0.1282	0.1138	0.0798	0.2400
	CA-RNN	**0.2028**	**0.4171**	**0.5216**	**0.2028**	**0.1390**	**0.0948**	**0.3074**
Fashion	POP	0.0024	0.0246	0.0272	0.0024	0.0082	0.0049	0.0314
	BPR	0.0145	0.0462	0.0783	0.0145	0.0154	0.0142	0.0510
	FM	0.0241	0.0660	0.1070	0.0241	0.0220	0.0195	0.0732
	CARS2	0.0257	0.0687	0.1107	0.0257	0.0229	0.0201	0.0761
	FPMC	0.0581	0.1869	0.2776	0.0581	0.0623	0.0505	0.1261
	HRM	0.0629	0.2024	0.2969	0.0629	0.0675	0.0540	0.1366
	RNN	0.0859	0.2696	0.3849	0.0859	0.0899	0.0700	0.1828
	CA-RNN	**0.1648**	**0.4305**	**0.5804**	**0.1648**	**0.1435**	**0.1055**	**0.2932**

by 58.2%, 22.3%, 18.8%, and 28.1%, respectively, on the Tmall dataset. And on the Fashion datasets, the relative improvements become 91.8%, 59.7%, 50.8%, and 60.4%. These great improvements indicate the superiority of our method brought by modeling sequential information and contextual information.

6.4 Experimental Analysis

In this section, we analyze the HIR and CA-RNN methods in detail. We visualize the representations of entities, contexts, and their interaction obtained from HIR and carefully examine the relation between performance and interacting order of entities and contexts. For context-aware sequential prediction, we conduct experiments to compare impacts of input contexts and aggregation methods and also assess the effects of input contexts and transition contexts.

6.4.1 Representation Visualization

Here, using the Delicious dataset as an example, we plan to demonstrate the representations in HIR and describe some interesting observations. In Fig. 6.2, we use Principal Component Analysis (PCA) to project the representations into a two-dimensional space. The distance in this figure reveals the relation of different representations.

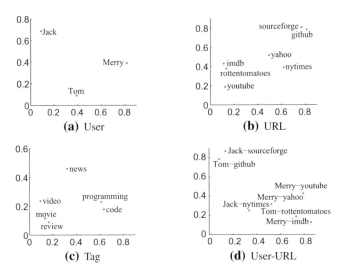

Fig. 6.2 Visualization of interaction representation in the Delicious dataset

Fig. 6.3 Visualization of user-URL-tag interaction representation in the Delicious dataset

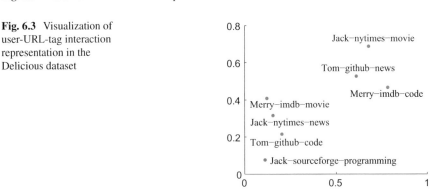

We select three distinguishable users and name them as *Jack*, *Tom*, and *Marry*, as shown in Fig. 6.2a. The representations of *URL* and *tag* are illustrated in Fig. 6.2b and c, respectively. We can observe that the representations of similar *URLs* or *tags* are close. Figure 6.2d shows the joint representations of *user* and *URL*. Although *Jack* and *Tom* have different representations, they have similar interaction representations when visiting programming Web sites. Similarly, although *Tom* and *Marry* have different representations, they have similar interaction representations when visiting movie Web sites. Figure 6.3 illustrates the interaction representations of *user*, *URL*, and *tag*. We can observe that positive samples and negative samples in the dataset are divided into two groups in the two-dimensional space according to their interaction representations.

6.4.2   Impact of Interacting Order

For some applications, the orders of entities and contexts are sometimes difficult to be determined exactly. In such situations, it will cost lots of labor to determine the proper orders. Thus, to examine the impact of interacting order of entities in HIR, we conduct experiments with all possible orders for two tasks: context-aware recommendation on the Food dataset and latent collaborative retrieval on the Delicious dataset. We also implement HIR with 0, 1, and 2 hidden layers in this experiment.

The RMSE results under different interacting order on the Food dataset is shown in Table 6.7. The variances of RMSE results of HIR are also shown in this table. Note that these results are all based on the same initialization. With the same number of hidden layers, performances of HIR are approximately the same with different interacting orders. The small variances also show that the impact of the interacting order for context-aware recommendation is not significant. Similar conclusion that the impact of the interacting order for latent collaborative retrieval is not significant can also be drawn from Table 6.8. In a word, in these tasks, different interacting orders often achieve similar performances. The unnecessary of the settled structure can greatly reduce the labor for analyzing the raw data and selecting a suitable order when there is no obvious one.

Table 6.7   Performance of context-aware recommendation on the Food dataset, evaluated by RMSE under different interacting order of entities and contexts

Order	HIR	HIR+	HIR++
(user, context, item)	0.9757	0.9643	0.9586
(item, context, user)	0.9763	0.9652	0.9583
(user, item, context)	0.9744	0.9633	0.9582
(context, user, item)	0.9786	0.9655	0.9596
(item, user, context)	0.9782	0.9663	0.9603
(context, item, user)	0.9798	0.9687	0.9594
Variance	4.12E-06	3.45E-06	6.95E-07

Table 6.8   Performance of latent collaborative retrieval on the Delicious dataset, evaluated by MAP under different interacting order of entities and contexts

Order	HIR	HIR+	HIR++
(user, URL, tag)	0.1222	0.1255	0.1266
(tag, URL, user)	0.1243	0.1263	0.1275
(user, tag, URL)	0.1196	0.1234	0.1243
(URL, user,tag)	0.1217	0.1242	0.1262
(tag, user, URL)	0.1232	0.1251	0.1267
(URL, tag, user)	0.1206	0.1228	0.1231
Variance	2.91E-06	1.75E-06	2.80E-06

Table 6.9 Performance comparison of different types of input contexts and different aggregation methods on the Tmall and Fashion datasets with dimensionality $d = 10$

	Method	Recall@1	Recall@5	Recall@10	F1-score@1	F1-score@5	F1-score@10	MAP
Tmall	CA-RNN-week	0.1681	0.3788	0.4984	0.1681	0.1263	0.0906	0.2793
	CA-RNN-month	0.1732	0.3688	0.4878	0.1732	0.1229	0.0887	0.2752
	CA-RNN-lin	**0.1748**	**0.4266**	**0.5515**	**0.1748**	**0.1422**	**0.1003**	**0.2986**
	CA-RNN-mul	0.1713	0.4068	0.5222	0.1713	0.1356	0.0950	0.2901
	CA-RNN-com	0.1765	0.3934	0.5068	0.1765	0.1312	0.0921	0.2865
Fashion	CA-RNN-week	0.1087	0.3325	0.4713	0.1087	0.1108	0.0857	0.2237
	CA-RNN-month	0.0839	0.2793	0.4125	0.0839	0.0931	0.0750	0.1858
	CA-RNN-lin	**0.1324**	**0.3720**	**0.5281**	**0.1324**	**0.1240**	**0.0960**	**0.2556**
	CA-RNN-mul	0.1157	0.3155	0.4861	0.1157	0.1052	0.0884	0.2298
	CA-RNN-com	0.1206	0.3376	0.4914	0.1206	0.1125	0.0893	0.2346

6.4.3 Analysis of Input Contexts

We conduct experiments on comparing different input contexts and different aggregation methods of input contexts. To compare the performance with input contexts *week* and *month* and different compare aggregation methods, Table 6.9 illustrates performance comparison of CA-RNN-week, CA-RNN-month, CA-RNN-lin, CA-RNN-mul, and CA-RNN-com with dimensionality $d = 10$.

From the performance of CA-RNN-week and CA-RNN-month, we can observe that the effect of the context *week* is slightly greater than that of the context *month* on the Tmall dataset. But on the Fashion dataset, the difference is relatively large, and CA-RNN-month performs poorly. It indicates that different days in a week can significantly affect user behaviors. The context *month* has significant effects on which category to buy, e.g., clothes or necessities of life in the Tmall dataset, but has limited effects on which item to buy, e.g., a specific cloth or bag in the Fashion dataset.

Moreover, from the results of CA-RNN-lin, CA-RNN-mul, and CA-RNN-com, we can clearly observe that the performances with both types of input contexts are better than those with only one type. And CA-RNN-lin achieves the best performances, which indicates that the linear method is the best aggregation method on this data. Thus, on other experiments, we use the linear method for aggregation of *week* and *month*.

Table 6.10 Performance comparison between input contexts and transition contexts on the Tmall and Fashion datasets with dimensionality $d = 10$

	Method	Recall@1	Recall@5	Recall@10	F1-score@1	F1-score@5	F1-score@10	MAP
Tmall	CA-RNN-input	0.1748	**0.4266**	**0.5515**	0.1748	**0.1422**	**0.1003**	0.2986
	CA-RNN-transition	0.1887	0.4134	0.5158	0.1887	0.1378	0.0938	0.3020
	CA-RNN	**0.2028**	0.4171	0.5216	**0.2028**	0.1390	0.0948	**0.3074**
Fashion	CA-RNN-input	0.1324	0.3720	0.5281	0.1324	0.1240	0.0960	0.2556
	CA-RNN-transition	0.1129	0.3284	0.4392	0.1129	0.1095	0.0799	0.2186
	CA-RNN	**0.1648**	**0.4305**	**0.5804**	**0.1648**	**0.1435**	**0.1055**	**0.2932**

6.4.4 Input Contexts Versus Transition Contexts

To compare input contexts and transition contexts, we illustrate the performances of CA-RNN-input, CA-RNN-transition, and CA-RNN with dimensionality $d = 10$ in Table 6.10.

On the Tmall dataset, the performance of CA-RNN-transition is very close to CA-RNN-input. But on the Fashion dataset, CA-RNN-input outperforms CA-RNN-transition a lot. Though transition contexts can bring performance improvement, the results may indicate that the effects of transition contexts are not very stable. It has limited effects on predicting user behaviors on purchasing fashion products, such as clothes, bags. Moreover, the results also indicate that the contextual information in input contexts and transition contexts has distinct effects on predicting future behaviors. Thus, it is necessary to incorporate both kinds of contexts to better predict the future. Incorporating both input contexts and transition contexts in one framework, CA-RNN achieves the best performance among our proposed methods. And it can significantly outperform models with only input contexts or transition contexts.

6.5 Conclusions

This chapter assesses the performance of COT, HIR and CA-RNN and other methods on different collaborative prediction tasks. These tasks are general recommendation, context-aware recommendation, latent collaborative retrieval, and click-through rate prediction. The empirical results indicate that COT, HIR, and CA-RNN outperform the state-of-the-art methods of context-aware collaborative prediction and sequential prediction. In addition, there are some interesting observations when visualizing the representations of entities, contexts, and their interaction. The performances with different interacting orders of entities and contexts are examined. On context-aware sequential prediction, experiments are conducted on comparing performance with different input contexts and different aggregation methods of input contexts. Besides, the performance of models is assessed with only input contexts and transition contexts.

References

1. Wu, S., Liu, Q., Wang, L., Tan, T.: Contextual operation for recommender systems. IEEE TKDE **28**, 2000–2012 (2016)
2. Liu, Q., Wu, S., Wang, L.: Collaborative prediction for multi-entity interaction with hierarchical representation. In: CIKM, pp. 613–622 (2015)
3. Liu, Q., Wu, S., Wang, L., Tan, T.: Predicting the next location: a recurrent model with spatial and temporal contexts. In: AAAI, pp. 194–200 (2016)
4. McMahan, H.B., Holt, G., Sculley, D., Young, M., Ebner, D., Grady, J., Nie, L., Phillips, T., Davydov, E., Golovin, D., et al.: Ad click prediction: a view from the trenches. In: Proceedings of the 19th ACM SIGKDD International Conference on Knowledge Discovery and Data Mining, pp. 1222–1230. ACM (2013)
5. Yan, L., Li, W.J., Xue, G.R., Han, D.: Coupled group lasso for web-scale ctr prediction in display advertising. In: Proceedings of the 31th International Conference on Machine Learning, pp. 802–810. ACM (2014)
6. Mnih, A., Salakhutdinov, R.: Probabilistic matrix factorization. In: Proceedings on Neural Information Processing Systems (2007)
7. Rendle, S., Freudenthaler, C., Gantner, Z., Schmidt-Thieme, L.: BPR: Bayesian personalized ranking from implicit feedback. In: Proceedings of the 25th Conference on Uncertainty in Artificial Intelligence, pp. 452–461. AUAI Press (2009)
8. Koren, Y.: Factorization meets the neighborhood: a multifaceted collaborative filtering model. In: Proceedings of the 14th ACM SIGKDD International Conference on Knowledge Discovery and Data Mining, pp. 426–434. ACM (2008)
9. Rendle, S.: Factorization machines. In: Proceedings of the 10th IEEE International Conference on Data Mining, pp. 995–1000. IEEE (2010)
10. Xiong, L., Chen, X., Huang, T.K., Schneider, J.G., Carbonell, J.G.: Temporal collaborative filtering with bayesian probabilistic tensor factorization. In: Proceedings of the 2010 SIAM International Conference on Data Mining, pp. 211–222. SIAM (2010)
11. Karatzoglou, A., Amatriain, X., Baltrunas, L., Oliver, N.: Multiverse recommendation: n-dimensional tensor factorization for context-aware collaborative filtering. In: Proceedings of the Fourth ACM conference on Recommender Systems, pp. 79–86. ACM (2010)
12. Jamali, M., Lakshmanan, L.: Heteromf: recommendation in heterogeneous information networks using context dependent factor models. In: Proceedings of the 22nd International Conference on World Wide Web, International World Wide Web Conferences Steering Committee, pp. 643–654 (2013)
13. Shi, Y., Karatzoglou, A., Baltrunas, L., Larson, M., Hanjalic, A.: Cars2: Learning context-aware representations for context-aware recommendations. In: Proceedings of the 23rd ACM International Conference on Information and Knowledge Management, pp. 291–300. ACM (2014)
14. Weston, J., Wang, C., Weiss, R., Berenzweig, A.: Latent collaborative retrieval. In: Proceedings of the 29th International Conference on Machine Learning, pp. 9–16. ACM (2012)
15. Rendle, S., Freudenthaler, C., Schmidt-Thieme, L.: Factorizing personalized Markov chains for next-basket recommendation. In: WWW, pp. 811–820 (2010)
16. Wang, P., Guo, J., Lan, Y., Xu, J., Wan, S., Cheng, X.: Learning hierarchical representation model for next basket recommendation. In: SIGIR, pp. 403–412. ACM (2015)

17. Zhang, Y., Dai, H., Xu, C., Feng, J., Wang, T., Bian, J., Wang, B., Liu, T.Y.: Sequential click prediction for sponsored search with recurrent neural networks. In: AAAI, pp. 1369–1376 (2014)
18. Ono, C., Takishima, Y., Motomura, Y., Asoh, H.: Context-aware preference model based on a study of difference between real and supposed situation data. In: User Modeling, Adaptation, and Personalization, pp. 102–113. Springer (2009)
19. Adomavicius, G., Sankaranarayanan, R., Sen, S., Tuzhilin, A.: Incorporating contextual information in recommender systems using a multidimensional approach. ACM Trans. Inf. Syst. (TOIS) 23(1), 103–145 (2005)

Printed in the United States
By Bookmasters